SEISMIC AND SEQUENCE STRATIGRAPHY AND INTEGRATED STRATIGRAPHY - NEW INSIGHTS AND CONTRIBUTIONS

Edited by **Gemma Aiello**

Seismic and Sequence Stratigraphy and Integrated Stratigraphy - New Insights and Contributions
http://dx.doi.org/10.5772/66242
Edited by Gemma Aiello

Contributors

Nickolay Ivanovich Akulov, Andrey Frolov, Irina Mashchuk, Shu Jiang, Wen Lin, Zhenlin Chen, Lei Chen, Lanyu Wu, Wei Dang, Yajun Li, Yue Wu, Dazhong Dong, Caineng Zou, Jinliang Zhang, Jingzhe Li, Pavel Kabanov, Jih-Hsin Chang, Ho-Han Hsu, Chih-Chieh Su, Char-Shine Liu, Shye-Donq Chiu, Yu-Fang Ma, Yuan-Wei Li, Yen-Chun Lin, Jen-Sen Shen, Eason Yi-Cheng Yang, Claudia Galli, Ricardo Alonso, Beatriz Coira, Gemma Aiello

Published by InTech
Janeza Trdine 9, 51000 Rijeka, Croatia

Publishing Process Manager Martina Usljebrka
Technical Editor SPi Global
Cover InTech Design team

First published December, 2017
Printed in Croatia
Legal deposit, Croatia: National and University Library in Zagreb

Additional hard copies can be obtained from orders@intechopen.com

Seismic and Sequence Stratigraphy and Integrated Stratigraphy - New Insights and Contributions, Edited by Gemma Aiello
p. cm.
Print ISBN 978-953-51-3675-0
Online ISBN 978-953-51-3676-7

For EU product safety concerns:
IN TECH d.o.o., Prolaz Marije Krucifikse Kozulić 3, 51000 Rijeka, Croatia,
info@intechopen.com or visit our website at intechopen.com.

Meet the editor

Dr. Gemma Aiello was born in Aversa (CE), Italy, on 24 October 1964. In 1989, she graduated in Geological Sciences at the University of Naples "Federico II." In 1993, she earned a PhD degree in Sedimentary Geology at the University of Naples "Federico II," Department of Earth Sciences, Faculty of Geological Sciences. She completed a 2-year postdoctoral fellowship at the University of Naples "Federico II," a CNR-CEE fellowship, and several contracts at the Research Institute "Geomare Sud," CNR, Naples, Italy. Since 1998, she has been a full-time researcher at the Italian CNR. Dr. Aiello has a 25-year experience in the field of sedimentary geology, marine geology, and geophysics, participating in different research projects of Italian National Research Council (CARG, Vector, Centri Regionali di Competenza). She was a contract professor of sedimentology and stratigraphy at the Parthenope University of Naples, Italy, and a teacher in formation courses of technicians in marine science and engineering in Naples, Italy.

Contents

Preface

This book contains seven chapters dealing with the investigation of seismic and sequence stratigraphy and integrated stratigraphy, including the stratigraphic unconformities, in different geological settings and using several techniques and methods, including the seismostratigraphic and the sequence stratigraphic analysis, the field geological survey, the well log stratigraphic interpretation, and the lithologic and paleobotanical data. Book chapters are separated into two main sections: (i) seismic and sequence stratigraphy and (ii) integrated stratigraphy. There are three chapters in the first section, including the application of sequence and seismic stratigraphy to the fine-grained shales, to the fluvial facies and depositional environments, and to the Late Miocene geological structures offshore of Taiwan. In the second section, there are three chapters dealing with the integrated stratigraphic investigation of Jurassic deposits of the southern Siberian platform, with the stratigraphic unconformities, reviewing the related geological concepts and studying examples from Middle-Upper Paleozoic successions, and, finally, with the integrated stratigraphy of the Cenozoic deposits of the Andean foreland basin (northwestern Argentina). The chapters separated into two main sections are preceded by an introductory chapter, namely the chapter 1, giving insights into the subject area.

Introductory chapter

The first chapter "An Introduction to the Seismic and Sequence Stratigraphy and to the Integrated Stratigraphy: Concepts and Meanings" by Gemma Aiello introduces the concepts of seismic and sequence stratigraphy and of integrated stratigraphy.In this chapter it is clarified that the sequence stratigraphic studies on the alluvial depositional systems and of the fine-grained shales, shown in this book, are of great actuality in a geological sense, mainly taking into account that the seismo-stratigraphic concepts have been developed on the marine deposits of passive-type Atlantic continental margins. The geologic evolution of the passive margins has been briefly introduced. Due to different case histories of this book located in China, the geological structure of the Chinese-type basins has been recalled as mainly controlled by the amalgamation of continental blocks, Paleozoic in age, associated with different insular arcs.

Seismic and Sequence Stratigraphy

The second chapter, "Sequence Stratigraphy of Fine-Grained Shale Deposits: Case Studies of Representative Shales in the USA and China," by Shu Jiang, Caineng Zou, Wen Lin, Zhenlin Chen, Lei Chen, Lanyu Wu, Wei Dang, and Yajun Li, discusses the importance of sequence stratigraphic studies of fine-grained shale deposits, hosting important oil and gas resources. This chapter points out the main vertical and lateral lithofacies variations and the patterns of well logs, coupled with geochemical and mineralogical data in order to reconstruct the sequence stratigraphic framework of significant shales located in the USA (Barnett shale, Woodford shale, Marcellus shale, Mowry shale, Niobrara shale) and in China (Longmaxi shale, Chang7 shale). The techniques and methods have been shown to aim at recognizing the sequence boundaries, the system tracts, the flooding surfaces, the sets of parasequences, and the parasequences for the reconstruction of the stratigraphic architecture of shales and of the corresponding control factors, including the relative sea-level fluctuations and the sedimentary supply.

The third chapter, "Sequence Stratigraphy of Fluvial Facies: A New Type of Representative from Wenliu Area, Bohai Bay Basin, China," by Jingzhe Li and Jinliang Zhang, deals with the application of sequence stratigraphic concepts to alluvial environments, as an alternative to the classical concepts of sequence stratigraphy, which have been typically developed on the marine environments. The Wenliu area (Bohai Basin, China), developed as a case history in this chapter, hosts a kind of rivers, namely, the distributive fluvial systems (DFS), characterized by a radiation patterns of the channels coming from a vertex, by a broad convex-concave upward downfan, and by the intersection of the alluvial system located on the river. A sedimentary model of the Wenliu area is herein proposed, showing that the prevalent lithologies are mudstones, sandy mudstones, muddy siltstones, shaley siltstones, and siltstones. Fifty-eight short-term sea-level cycles have been recognized based on sequence stratigraphic setting, which has evidenced the occurrence of both rises and falls of the sea level.

Chapter 4, "Seismic Stratigraphic Features of the Late Miocene-Present Unconformities and Related Seismic Units, Northern Offshore of Taiwan," by Jih-Hsin Chang, Eason Yi-Cheng Yang, Ho-Han Hsu, Chih-Chieh Su, Char-Shine Liu, Shye-Donq Chiu, Yuan-Wei Li, Yen-Chun Lin, and Jen-Sen Shen, explains the seismostratigraphic features in the northern Taiwan offshore based on seismic profiles. In this chapter, significant seismic unconformities and related seismic units are discussed and interpreted. The regional unconformities are interpreted to be formed as a response to a mountain collapse and to the end of regional volcanic and tectonic activity, genetically related with the development of Late Neogene basins in the northern Taiwan offshore. The tectonic evolution and the corresponding implications for the post-collisional basins have also been reconstructed.

Integrated Stratigraphy

Chapter 5, "Stratigraphy of Jurassic Sediments of the Southern Siberian Platform (Russia) Studied through Lithologic and Paleobotanical Data," by A.O. Frolov, N.I. Akulov, and I.M. Mashchuk, presents the results of an integrated stratigraphic study of Jurassic deposits of the Irkutsk Coal Basin through the integration of lithologic and paleobotanical data. The main stratigraphic units of the Irkutsk basin (Cheremkhovskaya Formation, Prisayanskaya Formation, and Kudinskaya Formation) have been studied through the individuation of their lithologic characteristics. Moreover, two new fossil plant assemblages have been individuated and described, which were previously unknown. The stratigraphic correlation has been established based on paleobotanical data, correlating the Jurassic deposits of the Irkutsk basin with the deposits filling the basins located in western Siberia.

Chapter 6, "Stratigraphic Unconformities: Review of the Concept and Examples from the Middle-Upper Paleozoic," by Pavel Kabanov, deals mainly with the disconformities and their geological meaning in stratigraphy. The disconformities have been examined at a seismic resolution, pointing out that on the seismic sections they appear if the subaerial exposure has provoked the development of a significant relief or alternatively of incised fluvial channels, which may be identified at the scale of the seismic profile. The evaluation of hiatuses related to disconformities has also been carried out, suggesting that these stratigraphic surfaces are subzonal and do not show index fossils below and above the unconformity itself. The conceptual expression of disconformities is suggested, since they can occur in flood plain successions and in shallow marine successions, both as deposited under a wet climate controlled by an open stream drainage and as deposited in a dry climate and a karst environment. The drowning unconformities, typically occurring in carbonate platform environment, have also been studied in this chapter. The case histories include the Permian-Pennsylvanian successions of the Sverdrup Basin (Arctic Archipelago, Canada), the Carboniferous successions of the Moscow Basin (Russia), and the Lower-Middle Devonian successions of the Mackenzie Corridor (northwestern Canada).

Chapter 7, "Integrated Stratigraphy of Cenozoic Andean Foreland Basin (Northern Argentina)," by Claudia Ines Galli, Ricardo Narciso Alonso, and Lidia Beatriz Coira, provides an integrated stratigraphic study of the Cenozoic deposits filling the foreland basin of the Andean fold and thrust belt (northwestern Argentina). The relationships between sedimentary processes and tectonic setting have been evidenced, since the geological history of the foreland basin is strictly related with the tectonic control factor, acting on the development of the alluvial systems occurring in the basin. This integrated stratigraphic study has been carried out integrating different stratigraphic data and methods, including the paleoenvironmental characters, the types of stratigraphic contacts between the units, the geochronological data referred to the age of the units, and the paleomagnetic data, constraining the age of the units inferred from geochronology. The pre-Andean basement, underlying the foreland basin filling, has been individuated and described in detail. The facies, the stratigraphic architecture, and the sequence stratigraphy of the Los Colorados Formation, the Angastaco Formation, the Palo Pintado Formation, the San Felipe Formation, the Métan Subgroup, the Guanaco Formation, and the Piquete Formation have been described in detail.

I thank Mrs. Mirena Calmic and Mrs. Martina Usljebrka, Publishing Process Managers of InTech—Open Science, Open Minds—who have contributed to this book project on stratigraphy with competence and patience, following day after day the editorial activities and making possible the publishing of this book.

Dr. Gemma Aiello, PhD
Full-Time Researcher
National Research Council of Italy (CNR)
Institute of Marine and Coastal Environment (IAMC)
Naples, Italy

Section 1

Introduction

Chapter 1

Introductory Chapter: An Introduction to the Seismic and Sequence Stratigraphy and to the Integrated Stratigraphy: Concepts and Meanings

Gemma Aiello

Additional information is available at the end of the chapter

http://dx.doi.org/10.5772/intechopen.71678

1. Introduction

This is the introductory chapter of the book "Seismic and sequence stratigraphy and integrated stratigraphy - new insights and contributions." In this chapter, the research themes studied in this book have been introduced referring to the seismo-stratigraphic and sequence stratigraphic techniques and methodologies, pertaining, in particular, the fine-grained shales and the alluvial systems, the seismo-stratigraphic features of Late Miocene deposits offshore the northern Taiwan and to the integrated stratigraphic studies, including the stratigraphy of the Jurassic deposits in the Irkutsk sedimentary basin studied through lithologic and paleobotanical data, the disconformities in stratigraphy, reviewing their theoretical concepts and studying selected examples from Paleozoic successions and the integrated stratigraphy of the foreland basin of the Andean fold and thrust belt.

The topics and the research themes developed in this book are of the great actuality and should have a good impact on the scientific research community. In fact, the sequence stratigraphic studies and the seismo-stratigraphic concepts have been typically developed on the deposits pertaining to the marine environment in a geodynamic context of a passive, Atlantic-type continental margin [1–13]. In this book, instead, emphasis is given to the sequence stratigraphic studies performed on the alluvial systems and on the fine-grained shales.

The passive margins are characterized by thick successions of clastic and carbonate deposits, mainly of shallow water depositional environments, constituting sedimentary wedges thickening toward the ocean. The sedimentary wedge overlies a continental lithosphere segmented in horst and graben structures and tends to be prograding on the newly formed continental lithosphere. The sedimentary successions of a passive continental margin may reach thicknesses in the order of 14 kilometers and accumulate during and after the continental rifting and the formation of oceanic lithosphere. In a well-developed Atlantic-type continental margin, a continental shelf, continental slope and rise and basin occur [14–28].

Sequence stratigraphic interpretations of the alluvial systems and of the fine-grained shales, studied in this book, may be considered as both a counterpart and an integration of sequence stratigraphic analyses of marine deposits in continental shelf, slope and basin environments. The sequence stratigraphic setting of the fluvial depositional systems has been studied by several authors in different geological frameworks [29–35]. The distributive fluvial systems (DFS) [35], which have been investigated in this book, are a particular type of fluvial system, which is characterized by a downstream whose size decreases, is not bounded from valleys and shows a pattern of different rays coming from an apex. The sequence stratigraphy of the fine-grained shales is an interesting research topic of this book, and theoretical aspects applied to several geological settings have been pointed out by several papers [36–40]. In particular, the sequence stratigraphy of the Barnett Shale and subordinately of the Woodford Shale is among the most studied research topics regarding the shales and has been coupled with other geological methodologies, including the geochemistry and the evaluation of the gas content for petroleum studies [41–45].

In this book, different case studies located in China have been presented. To this aim, it should be useful to clarify the type of geological structure of the Chinese-type basins. The present-day geological setting of the Asia continent and, in particular, of China has been controlled by the amalgamation of several Paleozoic continental blocks and many insular arcs. Ziegler et al. [46] have attempted to follow the traces of the migration of some of these blocks up to their unification in the Laurasia continent of the Pangea. The paleogeographic reconstructions of the Chinese region at the end of the Paleozoic have allowed to distinguish three Precambrian platforms, which have been captured during the growth processes in the Paleozoic (i.e., the Tarim platforms, Northern China, and Southern China) [47]. During the formation of the Meso-Cenozoic megasuture belt, a new set of plates was produced, with the capture of the blocks of the Lut, Iran, Tibetan block, and Indochinese platform, merging with the initial Paleozoic nucleus. In the time interval ranging from the Upper Cretaceous to the Pliocene, the collision of the Asia with the Arabian block occurred, while during the Cenozoic, the collision of the Asia with the Indian block occurred [48]. The Chinese basins include the Ordos, Pre-Nan Shan, Tsaidam, Tarim, Turfan, and Dzungarian and are characterized by the occurrence of Mesozoic-Tertiary continental successions deformed by both strike-slip and reverse faults involving the Paleozoic basement [49]. Their individuation appears to be related to compressional stresses [50]. Three types of basins, hosting important oil and gas resources, have been distinguished in China, namely the extensional basins, the compressional basins, and the transitional basins [50]. The extensional basins prevail in the Eastern China, including the Songliao and Bohai Gulf basins (the second one has been investigated in this book). The compressional basins are mainly located in the Western China, including the Tarim and Junggar basins, while the transitional ones are mainly located in the Central China, including the Sichuan and Ordos basins [50].

In this book, another important research topic is represented by the Andean foreland basin, whose age is Cenozoic and located in northern Argentina. Its formation has been modeled from a geodynamic point of view, finding a coefficient of erosional transport of 3000 m^2/year and a coefficient of depositional transport of 20,000 m^2/year [51]. These parameters have predicted the basin geometry of the Andean foreland basin [51]. It is a belt of foreland basins located eastwards of the Central Andes and extending into different regions, including the eastern Bolivia, the northern Argentina, the Paraguay and the southwestern sector of Brazil. Perhaps, this belt is characterized by several depocenters, whose regional sediment distribution has been reconstructed by Horton and deCelles [52]. According to the obtained results [52], the sedimentary basin filling strongly varies in the depocentral zones including the wedge-top, the foredeep, the forebulge and the back-bulge depocentral zones. This variation is accompanied by significant variations in elevation and gravimetric anomalies [52]. Significant variations in the crustal thickness of Andes have been previously pointed out by geological studies [53]. Moreover, the lithospheric flexure of the central Andes and the corresponding bending of stratigraphic sequences have been investigated in detail [54].

2. Seismic stratigraphy

In this book, the seismo-stratigraphic setting of the Northern Taiwan offshore has been reconstructed based on the geologic interpretation of seismic sections (see Chapter 3). Moreover, in Chapter 5, significant results on the stratigraphic unconformities have been shown, focusing on examples of Paleozoic successions. Perhaps, it should be useful to clarify some seismo-stratigraphic concepts and methods.

While the stratigraphic analysis was previously based on the field geological survey, on the measurement of stratigraphic sections and on the lithologic and paleontologic descriptions, aimed at reconstructing the depositional environments and at correlating the stratigraphic sequences among them, this work methodology has been deeply changed after the onset of the seismic stratigraphy, which has allowed for obtaining detailed seismic records of the stratigraphic successions.

The approach to the seismic stratigraphy is based on the key concept that the seismic reflectors may be compared with the strata plans and, perhaps, the geometry of the seismic reflectors corresponds to the depositional geometry [55]. In this sense, the seismic stratigraphy represents a geological and geophysical approach to the stratigraphic analysis and interpretation.

The seismic reflectors occur in correspondence with significant contrasts of the acoustic impedance, which is a significant parameter in seismic stratigraphy. When an acoustic wave meets the interface separating two media having a different acoustic impedance, a part of the wave is transmitted to the other medium, while another part is reflected on the interface among the two media. The concept of the acoustic impedance allows for the calculation of the quantity of transmitted and reflected acoustic energy.

If we consider U as the energy of the wave crossing the media M_1 and M_2 and we suppose that Z_1 is the acoustic impedance of the medium M_1 and Z_2 is the acoustic impedance of the medium M_2, the transmitted energy U_t can be calculated through the following equation:

$$U_t = \frac{2\,Z_1}{Z_1 + Z_2} \cdot U \tag{1}$$

while the reflected energy U_r can be calculated through the following equation:

$$U_r = \frac{Z_2 - Z_1}{Z_1 + Z_2} \cdot U \tag{2}$$

The contrasts of acoustic impedance controlling the individuation of the seismic reflectors are located along surfaces corresponding to strata surfaces or to other discontinuities having a chronostratigraphic meaning. The strata surfaces represent the old surfaces of deposition, and then, they are coeval in the depositional area. The discontinuities are old erosional or non-depositional surfaces corresponding to significant stratigraphic gaps. Also if they represent events varying during the geological time, the discontinuities are considered as chronostratigraphic surfaces, since all the strata overlying the discontinuity are younger than the underlying strata [2, 9–12]. When identified on a seismic section, the discontinuities let to identify the most important lateral variations in the deposition of a stratigraphic succession. Moreover, they offer a geological basis in order to subdivide the stratigraphic successions in depositional sequences, which are the basic stratigraphic units of seismic stratigraphy [2, 9–12].

The main steps of the seismo-stratigraphic analysis are represented by the identification of the discontinuities and consequently of the depositional sequences, by the reconstruction of the original geometry of the sedimentary bodies and related sedimentary environments and by the chronostratigraphic correlation [2, 9–12].

The seismic sequence analysis allows for the identification of the depositional sequences. The geometric relationships between the lateral terminations of the strata and the discontinuities or the correlative conformities define the boundaries of the depositional sequences [2]. The lateral terminations of the strata with respect to the sequence boundaries individuate the configurations of onlap, downlap, continuity (lower boundaries) and of erosional truncation, toplap and continuity (upper boundaries) [1, 2, 7–12].

The seismic facies analysis deals with both the individuation and the geologic interpretation of the geometry, continuity, amplitude, frequency and velocity of the seismic reflectors, more than the outer shape of the sedimentary bodies and the seismic facies associations in a depositional sequence [2, 56–61]. In the modern development of this methodology, one aim is represented by the recognition of clusters or groups, representative of significant variations in the properties of the rocks, in the lithology and in the content of fluids. The cluster analysis offers a significant instrument in order to perform the classification of the shapes of the seismic traces grouping them into clusters, often using an unsupervised process without a previous definition of the clusters [57, 60, 61].

The analysis of relative sea-level fluctuations is based on the construction of chronostratigraphic diagrams and of curves of relative sea-level cycles [1, 6–10, 62]. In a chronostratigraphic section, reporting the chronological units in the ordinates of the graph, each layer has an equal time duration. Both erosional and non-depositional hiatuses may occur among the time surfaces corresponding to the layers of the depositional sequences. Three-dimensional

Wheeler chronostratigraphic diagrams represent a useful tool in the geological interpretation of the seismic sections [63–65]. While the conventional Wheeler diagrams, which are usually made by hand, include sketch diagrams showing the extent of chronostratigraphic sequences, new methods have been recently developed in order to construct a Wheeler diagram for a seismic three-dimensional volume [63–65].

3. Sequence stratigraphy

The concepts of depositional sequence, isochronous boundaries, and characteristic correlation geometry, which have been typically developed in the seismic stratigraphy, may be applied in the stratigraphic analysis of outcrops, representing, in that case, the sequence stratigraphy. Some beautiful examples of progradation, toplap, and other stratigraphic relationships have been described by Bosellini [66], an Italian geologist who has applied the concepts of the sequence stratigraphy to significant outcrops of the Triassic carbonate platforms of the Dolomites (Northern Italy). Bosellini [66] has described several types of progradational geometries occurring in spectacular outcrops located in the Dolomites at an outcrop scale comparable with one of the seismic sections. In the Dolomites, an episodic progradation of the carbonate platform has been suggested based on outcrop analysis. During the periods of high debris input, the progradation of the carbonate platform occurred, which was evidenced by the widening of shallow water carbonate depositional environments. On the contrary, during the periods of low debris input, the basinal sedimentation prevailed on the shallow water carbonate deposition. The onlap of the basinal facies at the toe of the carbonate slope may be observed in outcrop [66]. The progradational geometries have been interpreted accordingly with two different models, which have been named as two periods of Triassic times. In the Ladinian model, the progradation and the aggradation of the carbonate platform took place contemporaneously, indicating a phase of a relative sea-level rise. In the Carnian model, toplap geometries have been observed in the carbonate platform, indicating a phase of relative sea-level stand [66].

Numerous are the sequence stratigraphic studies carried out on carbonate platform outcrops. Stafleu and Schlager [67] have carried out a sequence stratigraphic study in which pseudo-toplap geometries have been identified in the Schlern and Raibl Formations. Prograding clinoforms have been identified in the Schlern Formation coupled with topset geometries [67]. Two lithological models have been constructed to explain the geologic evolution of the carbonate platform, that is, (1) rapid progradation of the carbonate platform coupled with slow aggradation and (2) toplap of the prograding clinoforms against the topsets deposited in the inner platform. The seismic models have generated a pseudo-toplap, which is not coincident with a toplap in the outcropping sections [67].

The siliciclastic sequence stratigraphy, its concepts and application have been resumed by Posamentier and Allen [68]. The key concepts of siliciclastic sequence stratigraphy have been considered, including the key stratigraphic surfaces, such as the transgressive surface, the maximum flooding surface, the ravinement surface and many others, and their geologic meaning.

The control factors on the deposition of sequences and system tracts have been considered, including the sea-level fluctuations, the sediment supply and the accommodation space [68].

Some applications of siliciclastic sequence stratigraphy have been given in the recognition of depositional sequences and system tracts from well logs coupled with seismic profiles and biostratigraphic data [69]. The integration of these stratigraphic methods has been applied to the Gulf of Mexico and has allowed for the prediction of reservoirs, seals and source rocks, useful in the petroleum exploration. The stratigraphic architecture has evidenced the occurrence of a complete depositional sequence, consisting of lowstand system tract (LST), transgressive system tract (TST), and highstand system tract (HST), whose stratigraphic signature has been identified based on well log interpretation. High-resolution paleobathymetric and biostratigraphic interpretation of well logs has detailed the general stratigraphic setting.

Some key concepts of sequence stratigraphy, particularly referring to the stratigraphic unconformities, are given in the Chapter 5 of this book. The stratigraphic unconformities are considered as main stratigraphic surfaces and their identification in outcrops can be constrained using the relative weathering maturity of the subaerial profile, the calibration through cyclostratigraphy, the absolute dating and the biostratigraphy. At the scale of the seismic profiles, the disconformities show concordant strata overlying and underlying the stratigraphic surface. In the sense of this chapter, they are considered to include the ravinement surfaces, which are important stratigraphic surfaces, related to the erosion during the transgressive movement of the landward margin of the transgressive system tract (TST) [70–72]. Moreover, the concept of drowning unconformity has been reviewed, considering this stratigraphic surface as one of the most important stratigraphic surfaces in carbonate platform settings [73–76]. These surfaces develop when the rate of vertical aggradation of the carbonate platform is lower than the rate of the accommodation space. Perhaps the deep water sedimentation tends to prevail on the shallow carbonate sedimentation, as evidenced by the individuation of the drowning unconformity. These kind of unconformities have been individuated offshore of the Apulian region in the Southern Adriatic Sea [62] and onshore in the Gargano Promontory, showing a well-developed carbonate platform margin-slope-basin succession [77].

4. Integrated stratigraphy

In this book, different studies on integrated stratigraphy have been presented, which are grouped in the second section of the book. These studies are based on the integration of several stratigraphic methodologies, including the lithologic and paleobotanical data, the three-dimensional seismic models, the biostratigraphy, the paleopedology and paleoaquifer studies, the lithologic logs of cores and their subaerial exposure profiles, the individuation of the eroded paleosols, the lithologic and lithofacies logs of wells and some corresponding measurements and, finally, the facies analysis aimed at individuating the depositional architecture and the sequence stratigraphic setting.

Different stratigraphic methods are involved in the integrated stratigraphy, including the chemostratigraphy, the isotopic stratigraphy, the oxygen isotopes, the carbon isotopes, the strontium isotopes, the orbital cyclostratigraphy, the response of the climate system to the orbital forcing, the orbital forcing and the sedimentary environments, the identification of cyclical features and the spectral analysis of time series. Particular attention must be given to the methods of absolute dating and to the geological timescale.

Numerous papers have been produced in the field of the integrated stratigraphy, covering a wide range of competences. The stratigraphic record of Gubbio (Central Apennines, Italy) is one of the most studied research topics in the integrated stratigraphy [78]. Cretaceous and Paleogene stratigraphy of the Central Apennines has been deeply studied from the beginning of 1900. A pioneer of these studies was Otto Renz (1906–1992). Many paleomagnetic investigations have also been carried out on the Mesozoic-Paleogene stratigraphic record of the Umbria–Marche basin. One of the most significant lithological types is represented by the Scaglia limestones, whose directions of remnant magnetization have indirectly given indications on the geodynamic evolution of the Adria African Promontory [79]. Starting from the Middle Jurassic, the magnetic stratigraphy of the Mesozoic-Paleogene succession of the Umbria–Marche basin has allowed to individuate a record of the geomagnetic polarity [78]. The value of this record has been confirmed from its correlation with the oceanic paleomagnetic records. The integrated stratigraphic studies of the Gubbio section include the individuation of an Early Cretaceous tectonic event in the Adria promontory, having insights from the Umbria-Marche pelagic basin, the Barremian-Aptian boundary in the Poggio Le Guaine core, giving evidence on the magnetic polarity chron M0r and on the oceanic anoxic event 1a, the *Rotalipora cushmani* extinction at Gubbio, a planktonic foraminifer testifying the emplacement of a large volcanic province and the evaluation of the environmental fluctuations during the late Cenomanian at Gubbio based on the ichnofabric [78].

In my opinion, another main research topic in the integrated stratigraphy is represented by the Messinian Global Stratotype Section and Point (GSSP) [80]. High-resolution integrated stratigraphy has been presented by Hilgen et al. [80], based on the integration of different stratigraphic methodologies, including the calcareous plankton biostratigraphy, the magnetic stratigraphy and the cyclic stratigraphy. The Messinian GSSP has been individuated at the base of the red layer of the cycle n. 15 in the section Oued Akrech. It coincides with the first occurrence of *Globorotalia mitumida* and is dated back at 7.251 My [80]. For the upper part of the Miocene, the Messinian is the standard chronostratigraphic unit, whose knowledge is due to the corresponding salinity crisis occurring in the Mediterranean Sea.

5. Outline of this book

Different stratigraphic studies have been carried out in this book. First, they include the sequence stratigraphic architecture of siliciclastic- and carbonate-dominated shales in USA and China, focusing on the implications in the reservoir prediction. The sequence stratigraphy of alluvial depositional environments has also been studied, defining a new type of fluvial facies, representative of the Bohai Bay Basin, which is located in Eastern China in extensional tectonic setting. In the northern Taiwan offshore, the main regional unconformities (U1 and U2) and the related seismic units (SU I, SU II, SU III) have been singled out as an answer to the collapse of the fold and thrust belt located in the emerged areas. A new stratigraphic scale for the Jurassic deposits of western Siberia has been constructed based on the correlation of these deposits with the surrounding regions. The theoretical aspects of the stratigraphic unconformities have been reviewed, focusing on the drowning unconformities (Middle Devonian drowning unconformity). The significance of this study is the integration among different aspects of stratigraphy. Most of the work which has been described in this book derives from detailed in situ observations and sophisticated stratigraphic analyses.

This book contains six chapters, as follows:

Chapter 2 (Sequence Stratigraphy of Fine-Grained "Shale" Deposits: Case Studies of Representative Shales in USA and China).

Chapter 3 (Sequence Stratigraphy of Fluvial Facies: A New Type Representative from Wenliu Area, Bohai Bay Basin, China).

Chapter 4 (Seismic Stratigraphic Features of the Late Miocene-Present Unconformities and Related Seismic Units, Northern Offshore Taiwan).

Chapter 5 (Stratigraphy of Jurassic Sediments of the Southern Siberian Platform (Russia) Studied Through Lithologic and Palaeobotanical Data).

Chapter 6 (Stratigraphic Unconformities: Review of the Concept and Examples from the Middle-Upper Paleozoic).

Chapter 7 (Integrated Stratigraphy of the Cenozoic Andean Foreland Basin (Northern Argentina)).

Author details

Gemma Aiello

Address all correspondence to: gemma.aiello@iamc.cnr.it

Institute of Marine and Coastal Environment (IAMC), National Research Council of Italy (CNR), Napoli, Italy

References

[1] Mitchum RM Jr. Seismic stratigraphy and global changes in sea level, Part II; Glossary of terms used in seismic stratigraphy. In: Payton CE, editor. Seismic Stratigraphy – Applications to Hydrocarbon Exploration. AAPG Memoir 26, 1977. p. 205-212

[2] Mitchum RM Jr, Vail PR, Sangree JB. Stratigraphic interpretation of seismic reflection patterns in depositional sequences. In: Payton CE, editor. Seismic Stratigraphy – Applications to Hydrocarbon Exploration. AAPG Memoir 26, 1977. p. 117-133

[3] Pitman WC. Relationship between eustasy and stratigraphic sequences of passive margins. GSA Bulletin. 1978;**89**:1389-1403

[4] Ricci Lucchi F, Colalongo ML, Cremonini G, Gaspari G, Iaccarino S, Papani G, Raffi S, Rio D. Evoluzione sedimentaria e paleogeografica nel margine appenninico. In: Cremonini G, Ricci Lucchi F, editors. Guida alla geologia del margine appenninico-padano, 17-46. Bologna, Italy: Guide Geologiche Regionali della Società Geologica Italiana; 1982

[5] Bott MHP. Subsidence mechanisms at passive continental margins. In: Watkins JS, Montadert L, Dickerson PW, editors. Geological and Geophysical Investigations of Continental Margins. AAPG Memoir 29. 1979. p. 3-10

[6] Hardenbol J, Vail PR, Ferrer J. Interpreting paleoenvironments, subsidence history and sea-level changes of passive margins from seismic and biostratigraphy. Oceanologica Acta, SP. 1981;33-34

[7] Vail PR. Sea Level Changes and Global Unconformities Seismic Sequence Interpretation: A Report of the JOIDES Subcommittee on the Future of Scientific Ocean Drilling. Woods Hole, March 7-8 1977.

[8] Vail PR, Todd RG. Northern North Sea Jurassic unconformities chronostratigraphy and sea-level changes from seismic stratigraphy. In: Illing LV, Hobson GD, editors. Petroleum Geology of the Continental Shelf of Northwest Europe. London: Heyden and Son Ltd.; 1981. p. 216-235

[9] Vail PR, Mitchum RM, Thompson S. Relative changes of sea level from coastal onlap. In: Payton CE, editor. Seismic Stratigraphy – Applications to Hydrocarbon Exploration. AAPG Memoir 26. 1977. p. 63-81

[10] Vail PR, Mitchum RM, Thompson S. Global cycles of relative change of sea level. In: Payton CE, editor. Seismic Stratigraphy – Applications to Hydrocarbon Exploration. AAPG Memoir 26. 1977. p. 83-97

[11] Vail PR, Todd RG, Sangree JB. Chronostratigraphic significance of seismic reflections. In: Payton CE, editor. Seismic Stratigraphy – Applications to Hydrocarbon Exploration. AAPG Memoir 26. 1977. p. 99-116

[12] Vail PR, Hardenbol J, Todd RG. Jurassic unconformities, chronostratigraphy and sea-level changes from seismic stratigraphy and biostratigraphy. In: Schlee JS, editor. Interregional Unconformities and Hydrocarbon Accumulation. AAPG Memoir 36. 1984. p. 129-144

[13] Watts AB, Steckler MS. Subsidence and eustasy at the continental margin of eastern North America. In: Talwani M, Hay WF, Ryan WBF, editors. Deep Drilling Results in the Atlantic ocean; Continental Margins and Paleoenvironment. American Geophysical Union, Maurice Ewing Series, 3. 1979. p. 218-234

[14] Sheridan RE, Grow JA. The Atlantic Continental Margin U.S1988. DOI: 10.1130/DNAG-GNA-I2. ISBN (electronic): 978-0-8137-5458-1

[15] Watkins JS, Montadert L, Dickerson PW. Geological and geophysical investigations of continental margins. AAPG Memoir. 1979;**29**:473

[16] Watts AB. The U.S. Atlantic continental margin: subsidence history, crustal structure and thermal evolution. In: Bally AW, editor. Geology of Passive Continental Margins: History, Structure and Sedimentologic Record. AAPG Education Course Note Series 19. 1981

[17] Bernoulli D, Jenkyins HC. Alpine, Mediterranean and Central Atlantic Mesozoic facies in relation to the early evolution of the Tethys. In: Dott RH, Shaver RH, editors. Modern and Ancient Geosynclinal Sedimentation. SEPM Special Publication 19. 1974. p. 129-160

[18] Blanchett R, Montadert L. Geology of continental margins. Oceanologica Acta. 293 pp

[19] Grow JA, Mattick RE, Schlee JS. Multichannel seismic depth sections and interval velocities over outer continental shelf and upper continental slope between Cape Hatteras and Cape Cod. In: Watkins JS, Montadert L, Dickerson PW, editors. Geological and Geophysical Investigations of Continental Margins. AAPG Mem. 29. 1979

[20] Grow JA, Hutchinson DR, Klitgord KD, Dillon WP, Schlee JS. Representative multichannel seismic profiles over the U.S. Atlantic margin. In: Bally AW, editor. Seismic Expression of Structural Styles – A picture and Work Atlas. AAPG Studies in Geology, 15. 1983

[21] Montadert L, Roberts DG, Auffret G, Bock W, Du Peuple PA, Hailwood A, Harrison W, Kagami H, Lumsden DN, Muller C, Schnitker D, Thompson TL, Timofeev PP. Rifting and subsidence on passive continental margins in the North East Atlantic. Nature. 1977;**268**:305-309

[22] Montadert L, Roberts DG, De Charpal O, Guennoc P. Rifting and subsidence of the northern continental margin of the Bay of Biscay. In: Initial Reports on the Deep Sea Drilling Project (DSDP), US Government Printing Office, Washington, 48. 1979. 1025-1060

[23] Montadert L, De Charpal O, Roberts DG, Guennoc P, Sibuet JC. Northeast Atlantic passive margins: rifting and subsidence process. In: Talwani M, Hay WW, Ryan WBF, Ewing M, Series 3, American Geophysical Union, 1979

[24] Watts AB. Models for the evolution of passive margins. Phanerozoic Rift Systems and Sedimentary Basins. Elsevier. 2012;33-57. DOI: 10.1016/B978-0-444-56356-9.00002-X

[25] Brun JP, Besslier MO. Mantle exhumation at passive margins. Earth and Planetary Science Letters. 1996;**142**:161-173

[26] Erickson SG. Sedimentary loading, lithospheric flexure, and subduction initiation at passive margins. Geology. 1993;**21**:125-128

[27] Leroy M, Dauteuil O, Cobbold PR. Incipient shortening of a passive margin: the mechanical roles of continental and oceanic lithospheres. Geophysical Journal International. 2004. DOI: 10.1111/j.1365-246X.2004.02400.x

[28] Lin AT, Watts AB, Hesselbo SP. Cenozoic stratigraphy and subsidence history of the South China Sea margin in the Taiwan region. Basin Research. 2003;**15**:453-478

[29] Wright P, Marriott SB. The sequence stratigraphy of fluvial depositional systems: the role of floodplain sediment storage. Sedimentary Geology. 1993;**86**:203-210

[30] Darlymple RW, Choi K. Morphologic and facies trends through the fluvial-marine transition in tide-dominated depositional systems: A schematic framework for environmental and sequence stratigraphic interpretation. Earth Science Reviews. 2007;**81**(3/4):135-174

[31] Wescott WA. Geomorphic thresholds and complex response of fluvial systems – some implications for sequence stratigraphy. AAPG Bulletin. 1993;**77**(7):1208-1218

[32] Miall AD. Architecture and sequence stratigraphy of Pleistocene fluvial systems in the Malay Basin based on seismic time-slice analysis. AAPG Bulletin. 2002;**86**(7)

[33] Catuneanu O. Principles of sequence stratigraphy. Elsevier: Amsterdam, The Netherlands; 386 p

[34] Bellotti P, Milli S, Tortora P, Valeri P. Physical stratigraphy and sedimentology of the Late Pleistocene-Holocene Tiber Delta depositional sequence. Sedimentology. 1995;**42**(4): 617-634

[35] Weissmann GS, Hartley AJ, Nichols GJ, Scuderi LA, Olson M, Buehler H, Banteah R. Fluvial form in modern continental sedimentary basins: Distributive fluvial systems. Geology. 2010;**38**(1):39-42

[36] Slatt RM, Abousleiman Y. Merging sequence stratigraphy and geomechanics for unconventional gas shales. The Leading Edge. 2011;**30**(3):274-282

[37] Aboulresh MO, Slatt RM. Lithofacies and sequence stratigraphy of the Barnett Shale in the east-central Fort Worth Basin, Texas. AAPG Bulletin. 2012;**96**(1):1-22

[38] Slatt RM, Rodriguez ND. Comparative sequence stratigraphy and organic geochemistry of gas shales: commonality or coincidence? Journal of Natural Gas Science and Engineering. 2012;**8**:68-84

[39] Algeo TJ, Schwark L, Chower J. High-resolution geochemistry and sequence stratigraphy of the Hushpuckney Shale (Swope Formation, eastern Kansas): implications for climato-environmental dynamics of the Late Pennsylvanian Midcontinent Seaway. Chemical Geology. 2004;**206**(3-4):259-288

[40] Armstrong HA, Turner BR, Makhlouf IM, Weedon GP, Williams M, Al Smadi A, Abu Salah A. Origin, sequence stratigraphy and depositional environment of an upper Ordovician (Hirnantian) deglacial black shale, Jordan. Palaeogeography, Palaeoclimatology, Palaeoecology. 2005;**220**(3-4):273-289

[41] Loucks RG, Ruppel SC. Mississippian Barnett Shale: lithofacies and depositional setting of a deep water shale gas succession in the Fort-Worth Basin, Texas. AAPG Bulletin. 2007;**91**(4):579-601

[42] Slatt RM, O'Brien NR. Pore types in the Barnett and Woodford gas shales: contribution to understanding gas storage and migration pathways in fine-grained rocks. AAPG Bulletin. 2011;**95**(12):2017-2030

[43] Singh P, Slatt RM, Coffey W. Barnett Shale – Unfolded: sedimentology, sequence stratigraphy and regional mapping. Gulf Coast Association of Geological SocietiesTrans actions. 2008;**58**:777-795

[44] Sarmiento M, Ducros M, Carpentier B, Lorant F, Cacas M, Fiornet S, Wolf S, Rohais S, Moretti I. Quantitative evaluation of TOC, organic porosity and gas retention distribution in a gas shale play using petroleum system modeling: application to the Mississippian Barnett Shale. Marine and Petroleum Geology. 2013;**45**:315-330

[45] Perez Altamar R, Marfurt K. Mineralogy-based brittleness prediction from surface seismic data: Application to the Barnett Shale. Interpretation. 2014;**2**(4):T255-T271

[46] Ziegler AM, Scotese CR, Johnson ME, Mc Kerrow WS, Bambach RK. Paleozoic biogeography of continents bordering on Iapetus (pre-Caledonian) and Rheic (pre-Hercynian). In: West RM, editor. Paleontology and Plate tectonics with Special Reference to the History of the Atlantic Ocean. Milwakee Public Museum Special Publications in Biology and Geology. 1977. Vol. 2. p. 1-23

[47] Dott RH, Batten RL. Evolution of the Earth. McGraw Hill: New York, USA; 1971. p. 649

[48] Molnar P, Tapponier P. Cenozoic tectonics of Asia: effects of a continental collision. Science. 1975;**189**:419-426

[49] Meyerhoff A, Willums J. O China: an oilman's look behind the Great Wall. International Petroleum Encyclopedia. 1978:413-419

[50] Desheng L. Basic characteristics of oil and gas basins in China. Journal of Asian Earth Sciences. 1996;(3-5):299-304

[51] Flemings PB, Jordan TEA. synthetic stratigraphic model of foreland basin development. Journal of Geophysical Research – Solid Earth. 1989;**10**:3851-3866

[52] Horton BK, DeCelles PG. The modern foreland basin system adjacent to the Central Andes. Geology. 1997;**25**(10):895-898

[53] Beck S, Zandt G, Myers SC, Wallace TC, Silver PG, Drake L. Crustal-thickness variations in the central Andes. Geology. 1996;**24**:407-410

[54] Watts AB, Lamb SH, Fairhead JD, Dewey JF. Lithospheric flexure and bending of central Andes. Earth and Planetary Science Letters. 1995;**134**:9-21

[55] Anstey NA. Simple Seismics. Boston: International Human Resources Development and Co.; 1982

[56] Dumay J, Fournier F. Multivariate statistical analyses applied to seismic facies recognition. Geophysics. 1988;**53**(9):1151-1159

[57] De Matos MC, Osorio P, Johann P. Unsupervised seismic facies analysis using wavelet transform and self-organizing maps. Geophysics. 2007;**72**(1):P9-P21

[58] West BP, May SR, Eastwood JE, Rossen C. Interactive seismic facies classification using textural attributes and neural networks. The Leading Edge. 2002;**21**(10):1042-1049

[59] Roksandic MM. Seismic facies analysis concepts. Geophysical Prospecting. 1978;**26**(2): 383-398

[60] Coleou T, Poupon M, Azbel K. Unsupervised seismic facies classification: A review and comparison of techniques and implementation. The Leading Edge. 2003;**22**(10):942-953

[61] Marroquin I, Brault J, Hart B. A visual data mining methodology for seismic facies analysis: Part I – Testing and comparison with other unsupervised clustering methods. Geophysics. 2009;**74**(1):P1-P11

[62] De Alteriis G, Aiello G. Stratigraphy and tectonics offshore of Puglia (Italy, Southern Adriatic sea). Marine Geology. 1993;**113**:197-212

[63] Stark T. Generation of a 3D seismic Wheeler Diagram from a high resolution Age Volume. SEG Technical Program Expanded Abstracts, 782-785

[64] de Bruin G, Hemstra N, Pouwel A. Stratigraphic surfaces in the depositional and chronostratigraphic (Wheeler-transformed) domain. The Leading Edge. 2007;**26**(7):883-886

[65] Qayyum F, de Groot P, Hemstra N. Using 3D Wheeler diagrams in seismic interpretation – the HorizonCube method. First Break. 2012;**30**(3):103-109

[66] Bosellini A. Progradation geometries of carbonate platforms: examples from the Triassic of the Dolomites, northern Italy. Sedimentology. 1984;**31**:1-24

[67] Stafleu J, Schlager W. Pseudo-toplap in seismic models of the Schlern-Raibl contact (Sella platform, northern Italy). Basin Research. 1993;**5**(1):55-65

[68] Posamentier HW, Allen GP. Siliciclastic sequence stratigraphy – concepts and applications. In: SEPM Concepts in Sedimentology and Paleontology Series, 7, 2000. 210 pp. ISBN 1-56576-070-0

[69] Mitchum RM, Sangree JB, Vail PR, Wornardt WW. Recognizing sequences and system tracts from well logs, seismic data and biostratigraphy: Examples from the Late Cenozoic of the Gulf of Mexico: Chapter 7: Recent Applications to Siliciclastic Sequence Stratigraphy. AAPG Special Volumes, 1993;**A169**:163-197

[70] Nummedal D, Swift D. J P Transgressive stratigraphy at sequence-bounding unconformities: some principles derived from Holocene and Cretaceous examples. Sea-level fluctuation and coastal evolution. 1987;**41**:241-260

[71] Catuneanu O. Sequence Stratigraphy: Guidelines for a Standard Methodology. Stratigraphy & Timescales. 2017. DOI: 10.1016/bs.sats.2017.07.003

[72] Zecchin M, Caffau M, Catuneanu O, Lenaz D. Discrimination between wave-ravinement surfaces and bedset boundaries in Pliocene shallow-marine deposits, Crotone Basin, southern Italy: An integrated sedimentological, micropalaeontological and mineralogical approach. Sedimentology. 2017. DOI: 10.1111/sed.12373

[73] Godet A. Drowning unconformities: palaeoenvironmental significance and involvement of global processes. Sedimentary Geology. 2013;**293**(1):45-66

[74] Schlager W. The paradox of drowned reefs and carbonate platforms. GSA Bulletin. 1981;**92**:197-211

[75] Schlager W. Drowning unconformities on carbonate platforms. In: Crevello PD, Wilson JL, Sarg JF, Read JS, editors. Controls on Carbonate Platform and Basin Development. SEPM Special Publication, 1989;**44**:15-25

[76] Schlager W, Camber O. Submarine slope angles, drowning unconformities and self-erosion of limestone escarpments. Geology. 1986;**14**:762-765

[77] Graziano R. Early Cretaceous drowning unconformities of the Apulia carbonate platform (Gargano Promontory, Southern Italy): Local fingerprints of global palaeoceanographic events. Terra Nova. 1999;**11**:245-250

[78] Menichetti M, Coccioni R, Montanari A. The Stratigraphic Record of Gubbio: Integrated Stratigraphy of the Late Cretaceous-Paleogene Umbria-Marche Pelagic Basin. GSA Special Papers, 524, https://dx.doi.org/10.1130/SPE524, ISBN print:9780813725246.

[79] Channell JET, D'Argenio B, Horvath F. Adria, the African Promontory, in Mesozoic Mediterranean palaeogeography. Earth Science Reviews. 1979;**15**(3):213-292

[80] Hilgen FJ, Bissoli L, Iaccarino S, Krijgsman W, Meijer R, Negri A, Villa G. Integrated stratigraphy and astrochronology of the Messinian GSSP at Oued Akrech (Atlantic Morocco). Earth and Planetary Science Letters. 2000;**182**:237-251

Section 2

Seismic and Sequence Stratigraphy

Chapter 2

Sequence Stratigraphy of Fine-Grained "Shale" Deposits: Case Studies of Representative Shales in the USA and China

Shu Jiang, Caineng Zou, Dazhong Dong, Wen Lin, Zhenlin Chen, Lei Chen, Lanyu Wu, Wei Dang and Yajun Li

Additional information is available at the end of the chapter

http://dx.doi.org/10.5772/intechopen.71137

Abstract

The fine-grained "shale" deposits host a vast amount of unconventional oil and gas resources. This chapter examines the variations in lithofacies, patterns of well logs, geochemistry, and mineralogy in order to construct a sequence stratigraphic framework of the representative marine Barnett, Woodford, Marcellus, Mowry, and Niobrara fine-grained "shales" (USA) and the marine Longmaxi shale and lacustrine Chang7 lacustrine shale (China). Practical methods are proposed in order to recognize the sequence boundaries, the flooding surfaces, the parasequences and parasequence sets, the system tracts, and variation patterns of facies and rock properties. The case studies for the sequence stratigraphy in the USA and China have revealed that the transgressive systems tract (TST) and the early highstand systems tract (EHST, if identifiable) of fine-grained "shales" have been deposited in anoxic settings. TST and EHST of the siliciclastic "shales" are characterized by high gamma ray, high TOC, and high quartz content, while TST and EHST of the carbonate-dominated fine-grained "shales" are characterized by low gamma ray, organic lean, and carbonate rich fine-grained deposits. The lithofacies, geochemistry, mineralogy, depositional evolution, and reservoir development have been predicted and correlated within a sequence stratigraphic framework for the suggested cases. The best reservoir with the best completion quality is developed in TST and HST in both siliciclastic-dominated and carbonate-dominated fine-grained "shales."

Keywords: sequence stratigraphy, shale, fine-grained, siliciclastic dominated, carbonate dominated

1. Introduction

The great success of shale gas exploration in North America has attracted increasing attention from petroleum geologists in the world. They are passionate about finding more commercial shale gas in other countries. Although the success of shale gas exploration in North America leads geologists to believe that shale gas exploration has a good prospect, how to find the sweet spot of shale gas in shale strata is still the main obstacle we need to face in the shale gas exploration and development. Sequence stratigraphy is a relatively new concept of stratigraphy and is a powerful tool to subdivide the stratigraphic intervals into geologically realistic and cyclic patterns and to predict the stratal patterns, the lithofacies variations, and the petroleum reservoirs within a chronostratigraphic unit. In the past 40 years, the sequence stratigraphic studies mainly focused on the coarse-grained siliciclastic and carbonate deposits and on the fine-grained turbiditic systems from lacustrine to deepwater settings, leading to a great success in the conventional petroleum exploration [1–5]. However, only limited studies on the high-frequency sequence and sedimentology have been conducted to study the sequence stratigraphy of shale strata deposited in deep water or restricted shallow water environment [6–8]. According to the traditional sequence stratigraphy, it is difficult to carry out a sequence stratigraphic analysis in relatively deep water areas where the shales are deposited [9]. This is due to the relatively homogenous and fine-grained nature of the shale interval [10] and to the conformity contact between upper and lower strata. Only newly acquired seismic data tied to geology can characterize the partial reservoir parameters, and it is very challenging to pick surfaces reflecting sea-level changes of fine-grained deposits on the seismic data if not tied by core and log data [11, 12]. Also, the GR log alone sometimes does not indicate high TOC of the maximum flooding surface within a sequence [13]. Nevertheless, some researchers have integrated the geochemistry, the geomechanics, the core and outcrop analysis, the seismic data, the well logs, the petrology, and the paleontology in order to perform a sequence stratigraphic analysis of the shales aimed at understanding the sequence boundaries, the major transgressive surface of erosion, the maximum flooding surface, the stacking patterns, the lithofacies variability, and the depositional models within the sequence [8–10, 14–25]. Some geologists have even tried to establish the criteria and models for sequence stratigraphy study in marine shale and predict the good reservoir potential shale gas vertically and laterally [8, 10, 12, 21, 22, 26]. So far, sequence stratigraphy of the shale is still poorly understood, and the traditional seismic stratigraphy approach to recognize the boundaries of sequences and system tracts does not work well for the fine-grained shale due to their relatively homogeneous fine-grained nature. Recent advances in property tests of the geochemistry, the mineralogy, the petrology, the petrophysics, and the geomechanics have provided the basis for the detailed characterization of the sequence stratigraphy of fine-grained shales. This chapter analyzes the spatial and temporal variations in the lithofacies, the well-logging responses, the geochemistry, the mineralogy, the hydrocarbon content and aims at developing the sequence stratigraphic frameworks for representative siliciclastic-dominated shales (marine Barnett, Woodford, Marcellus, and Mowry in the USA, Longmaxi and Chang7 shales in China) and carbonate-dominated shales (Niobrara, USA). The general variation trends of the shale properties have been analyzed within a sequence stratigraphic framework, so that they can be used to regionally correlate the shaly strata in a systematic manner and to identify, to predict, and to map the high-quality source rock and the productive fine-grained shale reservoirs.

http://dx.doi.org/10.5772/intechopen.71137

2. Data and methodology

The data for characterizing the sequence stratigraphy of fine-grained shales include the regional geologic data, the seismic data, the well logs, the lithofacies description in outcrops and cores, the organic geochemistry, the mineralogy, the thin sections, the SEMs, the QEMSCAN, and the geomechanics.

This chapter has employed a classic sequence stratigraphy model, consisting of the lowstand systems tract (LST), of the transgressive systems tract (TST), and of the highstand systems tract (HST) [2, 27]. The HST can be subdivided into early highstand systems tract (EHST) and late highstand systems tract (LHST) based on identifiable stacking patterns of parasequences and shale properties [24]. Slatt and Rodriguez [8] observed that many shales have cyclical patterns related to the eustatic sea-level fluctuations. They also stated that the general sequence stratigraphy of shales consists of a major transgressive surface of erosion or its correlative conformity and of an organic-rich TST bounded by a transgressive surface below and by a maximum flooding surface (MFS) or a condensed section (CS) above. The TST is then followed by upward decreasing gamma-ray trends within an interpreted HST (**Figure 1**, see appendix for the abbreviations used in this chapter). Typical time frames in which sea-level cycles (length of geologic time for a complete sea-level fluctuations) can be interpreted for shales (mainly based upon their fossil components) include second order, approximately 10–25 My; third order, approximately 1–3 My; fourth order, approximately 100,000–300,000 years [20].

For the methodology, this chapter adopts the following step-by-step workflow for the sequence stratigraphic analysis of fine-grained "shale" systems.

1. Study on regional geology;

2. Outcrop investigation on shale stratigraphy;

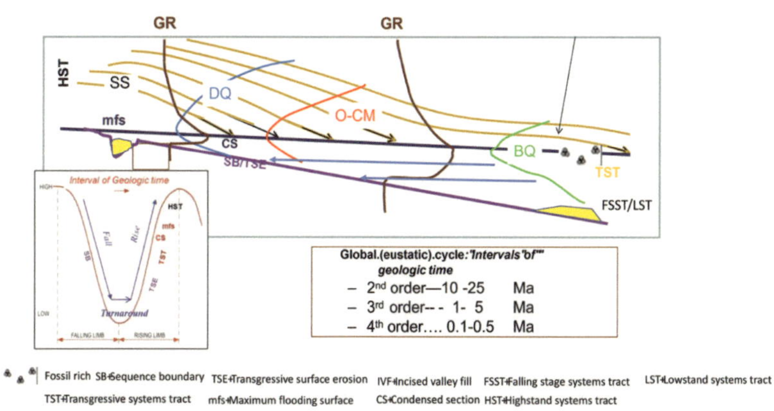

Figure 1. General sequence stratigraphic model of fine-grained shale. A conceptual gamma ray log is shown to indicate the log response of different fine-grained lithofacies (SS-shale rich in siltstone, DQ-shale rich in detrital quartz, O-CM-organic or clay rich mudstone, BQ-shale rich in biogenic quartz) in different stratigraphic interval within sequence stratigraphic framework. Modified from Slatt and Rodriguez (2012) and Slatt et al. (2014). See appendix for abbreviations of sequence stratigraphy from Van Wagoner et al. (1990).

3. Detailed lithological descriptions and preliminary interpretation of lithofacies from outcrop and cores;

4. Additional well logs and data collected from the cored wells were organized, including core-to-log depth corrections;

5. Lithofacies were identified based on core description, QEMSCAN analysis, thin section description, well log responses of spectral gamma ray, resistivity, FMI, density, and XRD data;

6. Seismo-stratigraphic, paleontological, geochemical, and mechanical data were used to identify the boundaries of different order sequences;

7. Vertical and lateral stacking patterns of the sequences were recognized;

8. High-frequency sequence stratigraphic framework was developed.

3. Lithofacies of shales

In the last few decades, the most of organic-rich shale-related plays, for example Barnett, Niobrara, Eagle Ford, Haynesville, and Bakken, have been labeled as "shale plays," and the terms "shale oil," "tight oil," and "resource play" are often used interchangeably in a public discourse [28–32]. The Barnett shale is a typical organic-rich siliceous shale (**Figure 2a**). The Eagle Ford shale is an organic-rich shale interbedded with thin fine-grained carbonate beds (**Figure 2b**). The lacustrine Green River shale consists of organic-rich shale with fine-grained ostracod rich carbonate (**Figure 2c**). The Niobrara shale play is dominated by a low permeability fine-grained chalk sourced by its adjacent marl source rock with continuous hydrocarbon accumulation [33]. The chalk and marl cannot be differentiated by naked eyes and appear like

Figure 2. Lithofacies of typical fine-grained "shales" in U.S. The Green River Fm is of lacustrine environment and the rest are of marine environment.

organic-rich argillaceous shale due to fine-grained nature and both containing organic matter (**Figure 2d** and **e**). The Bakken/Exshaw play (the USA and Canada) mostly consists of permeable fine-grained tight dolomite within upper and lower organic-rich shales (**Figure 2f**). The reason for which all these different lithofacies are called shale is the fine-grained nature of these deposits. Shale is defined as a fine-grained sedimentary rock that is a mix of clay minerals and quartz, calcite, and so on. The typical grain size range of shale is less than 1/256 of a millimeter in diameter. The ratio of clay to other minerals is variable. Shale used to be mistakenly considered as clay. For the organic-rich source rock interval, it usually has a mixed succession of interlayered fine-grained shales, sandstone, and/or carbonates. They are often called shale play due to their fine-grained grain size and to their close relationships with organic-rich shale. The shales representing different geologic settings in this study from different basins around the world refer to the fine-grained stratigraphic interval consisting of siliciclastic-dominated and carbonate-dominated lithofacies ranging from primary organic-rich shale to minor fine-grained organic-lean carbonate or sandstone interbeds.

4. Sequence stratigraphy of siliciclastic-dominated shales

Siliciclastic shales are mainly composed of silicate minerals and transported and deposited as fine-grained particles. The basic sequence stratigraphy of representative siliciclastic Barnett, Woodford, Marcellus, and Mowry shales is herein presented based on the variations in lithofacies, geochemistry, well log responses, and cyclicity of relative sea-level change and its effects upon stacking patterns and stratigraphy.

1. Barnett shale

The Mississippian Barnett shale is divided into lower and upper Barnett shale intervals separated in the northern part of the basin by the Forestburg Limestone. The lower Barnett sits directly above an SB/TSE, which caps the underlying Viola/Ellenburger limestones. The lower Barnett is dominated by siliceous mudstones and the upper Barnett is dominated by calcareous mudstones [8]. These lithofacies form distinctive stacking patterns termed "gamma-ray parasequences" (GRP) [34] and termed "high frequency-sequences" by Abouelresh and Slatt [21]. We classify the Lower Barnett shale and Forsberg limestone as a third-order sequence consisting of organic-rich high gamma-ray TST above the Ellenburger limestone and HST of generally decreasing gamma-ray interval. The high gamma-ray lower section of Upper Barnett shale represents TST of another third-order sequence (**Figure 3a**).

2. Woodford shale

The Late Devonian-Early Mississippian Woodford shale comprises three members. They are represented by the Lower Woodford Black shale, unconformably (SB/TSE) overlying (SB/TSE) the Hunton Group carbonates, by the Middle Woodford organic-rich black pyritic shale, and by the more quartzose-phosphatic Upper Woodford. The Woodford shale was deposited during a second-order sea-level cycle consisting of several third-order sequences [22]. Each third-order sequence comprises a TST with upward increasing in gamma ray and a HST with upward decreasing in gamma ray (**Figure 3b**).

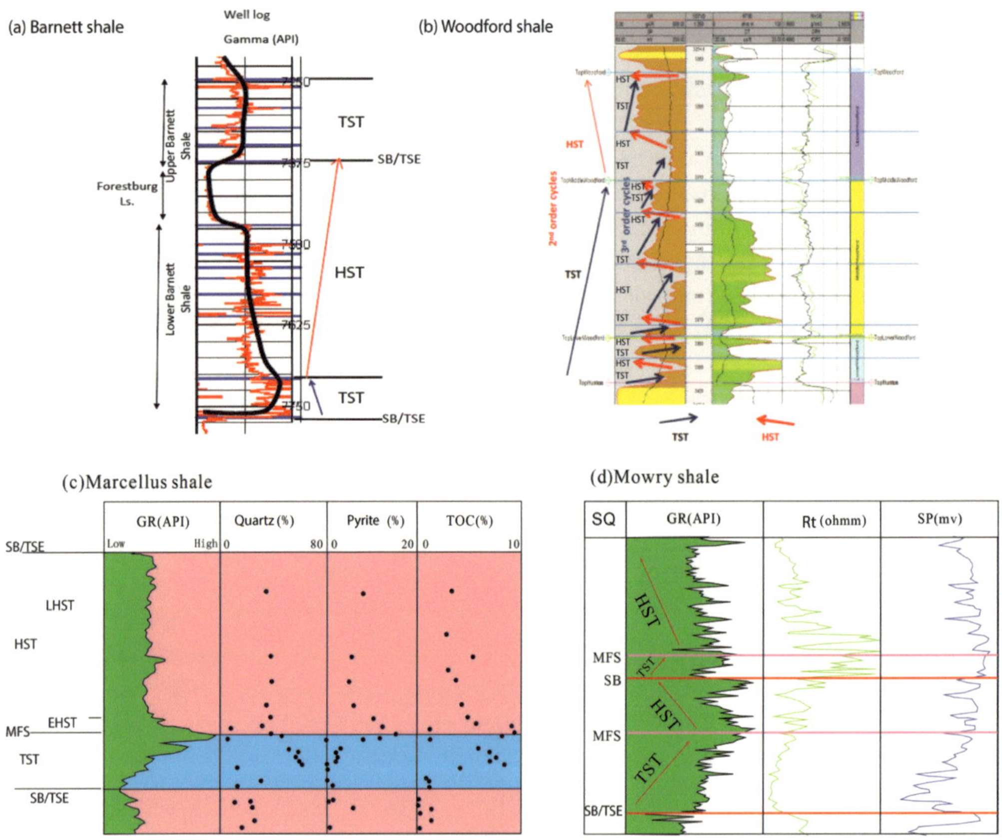

Figure 3. Sequence stratigraphy of typical Barnett, Woodord, Marcellus and Mowry siliciclastic marine shales in U.S. Original data of Barnett shale, Woodford shale, and Marcellus shale are from Slatt and Rodriguez (2012), Slatt and Rodriguez (2014), and Lash and Engelder (2011), respectively.

3. Marcellus shale

The organic-rich lower Union Springs Member of Devonian Marcellus Shale in the northeastern USA. was deposited as a third-order sequence in the Appalachian foreland basin [35]. This sequence consists of a lower, upward increasing API gamma-ray TST and an upper, upward decreasing API gamma-ray HST. The MFS constituting the highest gamma-ray shale separates the TST and HST, and the HST can further be divided into EHST (early stage) and LHST (late stage) (**Figure 3c**). Generally, the TST and EHST are rich in TOC and quartz content.

4. Mowry shale

The Cretaceous Mowry shale in Wyoming consists of third-order sequences with each consisting of TST and HST. TST sits sharply above the SB/TSE and is indicated by a shale interval of upward increasing in gamma ray and resistivity (R_t). The MFS has the highest gamma ray representing the highest sea level. HST generally consists of shale with upward decreasing in gamma ray and resistivity (R_t).

5. Longmaxi shale

The Lower Silurian Longmaxi shale with the best marine shale gas potential is a typical example. This shale is characterized by organic-rich marine shale with abundant graptolite [24]. The thicker organic-rich portion of this Silurian shale was mainly deposited in intra-shelf lows (bathymetric lows on the shelf) since the foredeep area was subject to clastic sediments dilution from the orogenic belt. The Lower Silurian Longmaxi shale was interpreted as a third-order sequence consisting of transgressive systems tract (TST) and highstand systems tract (HST). Based on the lithofacies and geochemical characteristics, the HST can be further subdivided into early highstand systems tract (EHST) and late highstand systems tract (LHST) (**Figure 4**). During the deposition of the lowstand systems tract (LST), wide shelf areas have experienced erosion, and the sediments have deposited in proximal intra-shelf lows. During the deposition of the transgressive systems tract (TST), the shoreline backstepped toward sediment source areas. The shale is generally deposited in the wide shelf, intra-shelf low, and slope area. The organic-rich shale interval with abundant in-situ graptolite, for example *Diplograptus* [24], high gamma-ray value, and high quartz content, is deposited in anoxic settings. During the deposition of the early highstand systems tract (EHST), the total organic content (TOC) in the shale may be high since the redox condition is still anoxic to dysoxic. The graptolite in EHST is characterized by transported microoffsite graptolite, for example *Monograptus* [24]. During

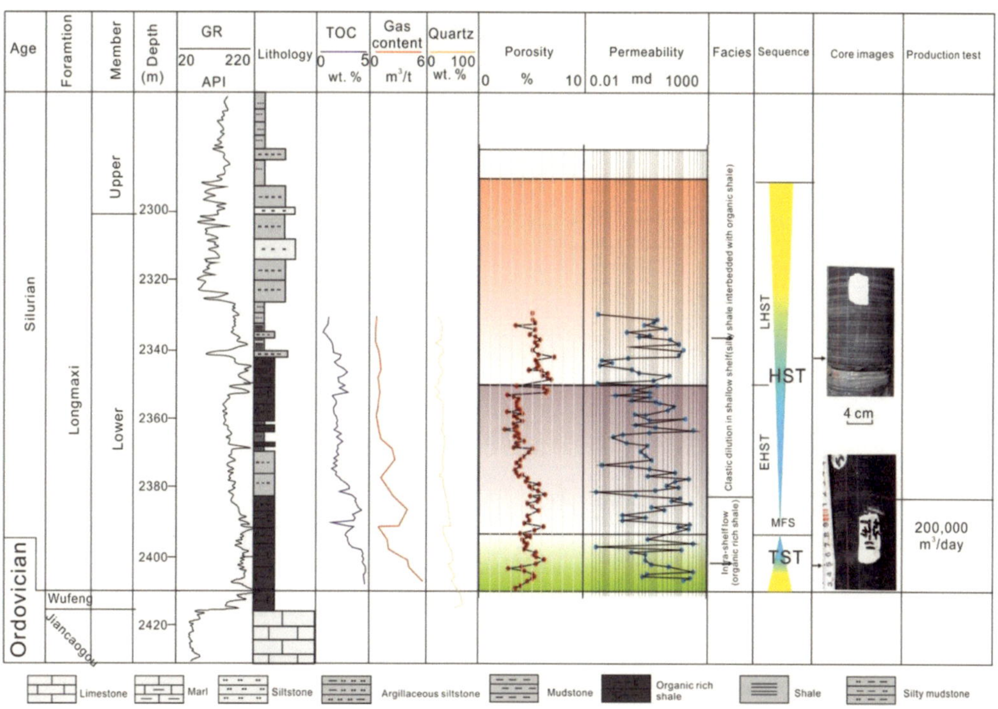

Figure 4. Sequence stratigraphy of Silurian siliciclastic dominated Longmaxi marine shale in Jiaoye1 well in SE Sichuan Basin, Southwest China.

the deposition of the late highstand systems tract (LHST), a further fall of sea level resulted in an increased input of siltstone and sandstone within the shales (see silty layer in core images of the LHST shale in **Figure 4**). During this time interval, sedimentary environment changes to oxygenated shallow marine setting and the coarsening-upward sequence made up of carbonatic shale and silty shales progrades basinward. The shale in LHST is organic-lean and clay-rich shale due to the dilution of clastic input. Generally, the Paleozoic marine shale in TST and EHST has high TOC, high quartz content, and high gas content and is the main target for shale gas play in China. The intervals with high TOC and quartz content in TST and EHST have higher porosity. The porosity in LHST is the highest due to the contribution of silty shale. The sandstone or carbonate in the LHST may trap gas migrated from the TST and EHST shale source rock, and they can form tight sand/carbonate gas play (**Figure 4**).

5. Sequence stratigraphy of carbonate-dominated shales

The sequence of carbonate-dominated shale has been studied using the Niobrara shale in Wyoming and Colorado (USA), for example. For the carbonate-dominated sequence, its sea-level response is different from that one of siliciclastic-dominated shales. For the siliciclastic shales, the sea-level rise has responses of increase in organic matter content and gamma ray (GR) of uranium. For the carbonate-dominated sequence, the sea-level rise means less input in siliciclastic clay and deposition of limestone in clean and quiet water. For the Niobrara equivalent formation in the Cretaceous Seaway, the sequence boundary is indicated by the maximum input of siliciclastic clay indicated by the highest GR in proximal Utah and clay-rich marl in Colorado (**Figure 5**). The TST is indicated by upward decrease in GR and deposit

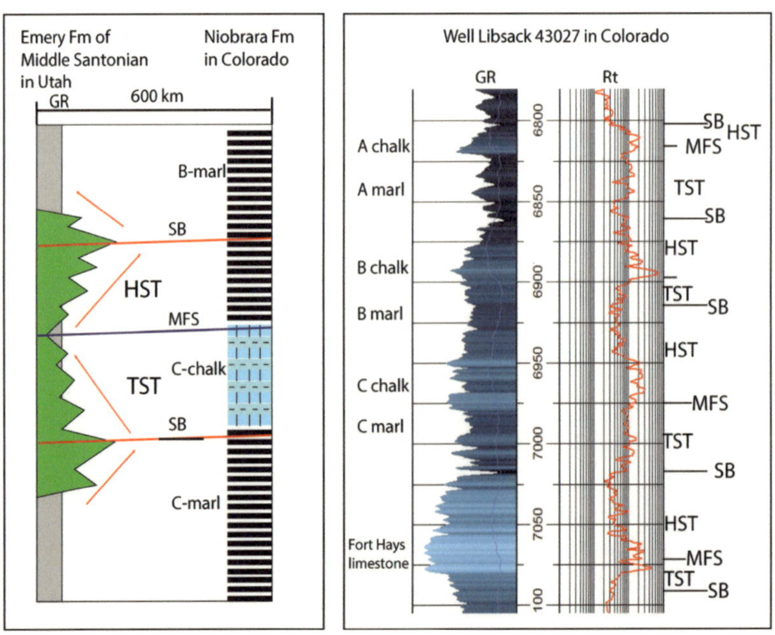

Figure 5. Sequence stratigraphy of carbonated dominated Niobrara shale in U.S.

of chalk. The HST is indicated by increase in GR and deposit of marl. The MFS is indicated by the lowest GR, the highest Rt, and clean chalk. The case study of Well Libsack 43,027 in Colorado in **Figure 5** clearly shows the sequence stratigraphic framework of Niobrara shale formation consisting of chalk and marl.

6. Sequence stratigraphy of fine-grained shale interval in lacustrine basin exampled by Ordos Basin in China

The late Cretaceous evolution of lakes and the corresponding depositional systems in the Ordos Basin of Northern China have been controlled by the changes in the lake level, the sediment supply, and the tectonic setting. The lakes expanded and shrank similarly to littoral to shallow marine environments [36]. Therefore, the sequence stratigraphy and system tracts developed in littoral to shallow marine environment can be applied in the lacustrine Ordos Basin [37] due to similarities of geologic processes in marine and lacustrine settings. To this aim, we have adopted methodology and terminologies of classical sequence stratigraphy also in lacustrine settings, such as the lowstand systems tract (LST), the transgressive systems tract (TST), the highstand systems tract (HST) [1, 2, 8].

For the shale play located in the Chang7 source rock of the Ordos Basin, we have used the first lacustrine shale gas well, LP177, for the shale play located in the Chang7 source rock of the Ordos Basin in order to reconstruct the lacustrine sequence stratigraphy. The Chang7 interval in well LP177 is characterized by organic-rich black shales overlain by gray organic-lean shales with sandstone interbeds (**Figure 6**). The SB is characterized by a change in depositional environments from semideep lake to shallow lake and by a corresponding change in lithofacies from thin siltstone within shale to thicker siltstones within shales. The LST is characterized by siltstone within shale. The transgressive surface bounding the LST and the TST has been identified through the sharp difference in color from gray shale to black shale and through the variations of the TOC content. The MFS is characterized by the highest gamma ray, the highest acoustic transit time (AC), high resistivity, and the highest TOC. HST is identified by coarsening-upward interval. The lower part of HST is characterized by organic-rich shale representing early highstand systems tract (EHST). The upper part of HST is characterized by silty interval representing later highstand systems tract (LHST). Within this sequence stratigraphic framework, the organic-rich shale having TOC of 2–6% was deposited during TST and EHST (**Figure 6**). Tests on five samples from the organic-rich Chang7 shale have indicated a porosity of 3–3.5% and a permeability of 60–100 nD. The 1.5–8 mg/g S1 in the shale reservoir interval indicates the accumulation of free oil in the shale or the presence of shale oil. The siltstone and sandstone intervals in LST and LHST are interpreted as turbidite deposits due to their position in semi to deep lake settings and their association with thick shale. GR curves and lithologies show sharp contact with the underlying and overlying shales confirming the nature of the turbidite deposits. Tests on the lower sandbody indicate a porosity of 11% and permeability of 0.1 mD (millidarcy), and it is classified as a tight oil play. The upper sandbody was tested to have a porosity of 14% and permeability of 1.3 D (darcy) and is classified as a conventional turbidite sandstone reservoir in large part due to the darcy-range permeability (**Figure 6**). The maturity of 0.77–0.95% of this organic-rich shale puts the Chang7 shale in the oil to wet gas window. The reported gas production of 2350 m^3/day (82,990 ft^3/

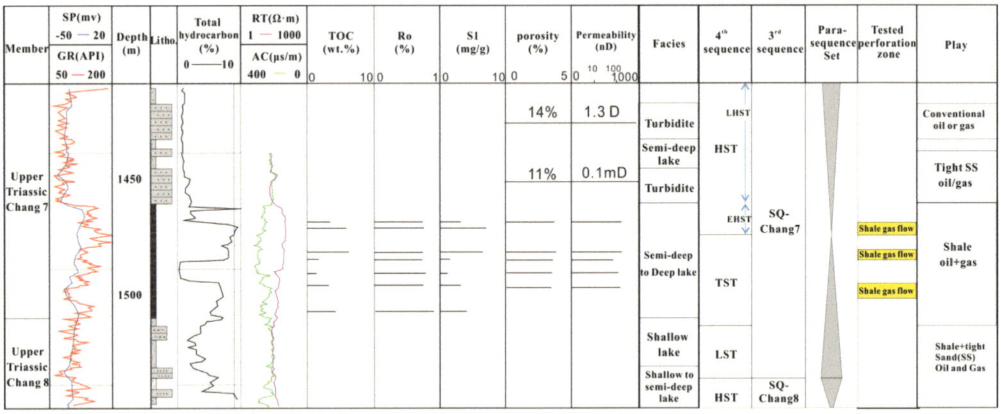

Figure 6. Sequence stratigraphy and its application of Triassic lacustrine Yanchang Formation, LP177 well in Ordos Basin, China.

day) from this organic-rich shale interval is probably wet shale gas indicated by the maturity. These results indicate that the LST and LHST within the sequence stratigraphic framework of lacustrine sediments tend to develop conventional and unconventional tight (sandstone) reservoirs, while TST and EHST are prone to develop shale gas and oil reservoirs.

7. Application

As shown above, an understanding of sequence stratigraphy can be a very powerful tool for building a sequence stratigraphic framework for the lithofacies prediction. It can also help to interpret the depositional history and to predict fine-grained unconventional resource plays for a single well to a basin (**Figure 3–6**). **Figure 4** shows the successful production of 200,000 m^3/day shale gas is from the predicted high-quality shale reservoir in TST and EHST interval with high TOC, high quartz content, and high gas content. At a regional scale, sequence stratigraphy can be used for the correlation of the stratigraphy, the lithology, the mineralogy, the geochemistry, and even rock fabrics. **Figure 7** shows that TST has the highest TOC content, total gas content and quartz content from Jiaoye1 well to Pengye1 well. At the same time, the TOC content has the similar changing patterns to gas content within Sequence A (lower member of the Longmaxi Fm): both show decreasing trends from TST to LHST. Also, gas content decreases with decreasing quartz content from TST to LHST, which indicates the shale in TST is high-quality fracturing susceptible shale gas reservoir with the highest TOC, gas content, and brittle minerals content (**Figure 7**). The positive relationship between TOC, quartz content, and shale gas content in the Silurian Longmaxi shale is similar to that one of the Bartnett shale in the USA, which is attributed to the formation of biogenic quartz in an anoxic environment that is favorable for organic matter preservation [24]. Laterally, shale gas content and quartz content in TST decreases from Jiaoye1 well to Pengye1 well even though they have similar TOC. These

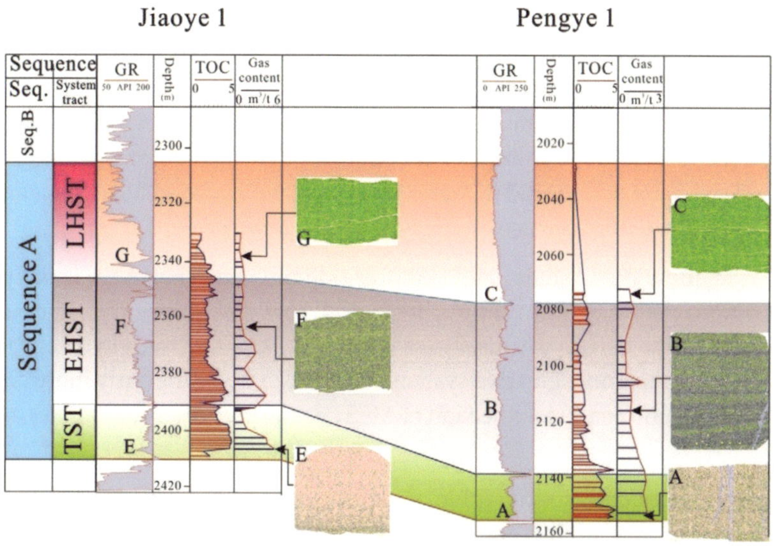

Figure 7. Correlation of stratigraphy, geochemistry, mineralogy, gas content, and rock fabrics within sequence stratigraphic framework across Jiaoye1 well and Pengye1 well. The light pink color represents quartz and the dark green color indicates clay minerals. The color indicates quartz content decreases upward and clay content increases upward at each well location.

verify the TST shale at the Jiaoye1 well has the better reservoir quality in time and space. The commercial development of the Jiaoshiba Shale Gas Field where Jiaoye1 is located proves the prediction of high-quality shale gas reservoir by using sequence stratigraphic approach.

8. Conclusions

The theory, approach, and terminology of clastic sequence stratigraphy developed for relatively coarse-grained deposits can be applied to fine-grained "shale" since their depositions are both controlled by sea- or lake-level changes. The sequence stratigraphic studies of marine and lacustrine shales in the USA and China have indicated that the sequence boundary of shales coincides with an unconformity formed during a relative sea-level or lake-level fall. The different and eventual occurrence of LST, TST, and HST has delineated the occurrence of both complete and/or incomplete depositional sequences, depending on their location. Many sequences of marine shales in the USA and China are incomplete, since they have developed only TST and HST. On the other hand, the lacustrine shale represented by the Chang7 shale in the Ordos Basin has developed complete depositional sequences, including LST, TST, and HST. HST can be further divided into EHST and LHST.

The investigation of siliciclastic-dominated fine-grained "shale" and carbonate-dominated fine-grained "shale" has revealed that lithofacies and well log responses to sea- and lake-level changes are strongly different. Previous proposals stated that high gamma ray and organic-rich shale are indicative of TST or the highest gamma ray and TOC are criteria for identifying MFS of siliciclastic-dominated fine-grained shales exampled by Barnett, Woodford, Marcellus, Mowry, and Longmaxi shales do not work for carbonate-dominated fine-grained shale. The sequence stratigraphy of the Niobrara carbonate-dominated fine-grained shale in this chapter shows that the rise in the sea level resulted in deposit of fine-grained low gamma-ray organic-lean chalky carbonate in clean water and the fall in the sea level resulted in clay-rich, high gamma ray, and organic-rich marl due to the increase in clay input during the sea-level drop.

Both siliciclastic-dominated "shale" and carbonate-dominated "shale" are usually located in anoxic depositional settings, for example, intra-shelf low and slope. The brittle minerals in rich siliciclastic marine and lacustrine shales with high TOC and high hydrocarbon (oil and gas) content occur in the transgressive systems tract (TST) up to the early highstand systems tract (EHST). The brittle condensed section of MFS has the best reservoir quality and completion quality.

Similarly to clastic sequence stratigraphic models developed for sandstones and carbonates, the sequence stratigraphy of fine-grained shale provides a powerful tool for predicting and mapping stratigraphy, variations of lithology, geochemistry, mineralogy, and shale gas content in time and space, and ultimately the most productive facies of unconventional gas shales.

Acknowledgements

This study was supported by EGI's multiyear China Shale Plays consortium (100980) and Open fund from Sinopec Key Laboratory of Shale Oil/Gas Exploration and Production Technology (Grant no. G5800-15-ZS-2X038), Key Laboratory of Tectonics and Petroleum Resources of Ministry of Education (China University of Geosciences) (TPR-2017-02), Key laboratory of strategy evaluation for Shale gas of Ministry of Land and Resources, U.S. National Science Foundation (No. 1661733) and China National Natural Science Foundation (nos. 41728004, 41302089 and 41472112). We are also grateful to reviewers and handling editor for their critical comments and constructive suggestions. The international collaboration between PetroChina, Sinopec, CNOOC, China University of Geosciences at Beijing, China University of Geosciences at Wuhan, China Geological Survey, China University of Petroleum at Beijing, China University of Petroleum at Qingdao and EGI also made this research and book chapter possible.

A. Appendix

Abbreviations for sequence stratigraphy used throughout this chapter include: SB, sequence boundary; TSE, transgressive surface of erosion; RSE, regressive surface of erosion; LST, lowstand (falling stage) systems tract; TST, transgressive systems tract; HST, highstand systems tract (consisting of EHST—early highstand systems tract and LHST—late highstand systems tract); CS, condensed section; MFS, maximum flooding surface; MRS, maximum regressive surface; TOC, total organic content (wt.%).

Author details

Shu Jiang[1,2,3,4,5]*, Caineng Zou[6], Dazhong Dong[6], Wen Lin[6], Zhenlin Chen[7], Lei Chen[8], Lanyu Wu[7], Wei Dang[9] and Yajun Li[10]

*Address all correspondence to: sjiang@egi.utah.edu

1 Research Institute of Unconventional Oil & Gas and Renewable Energy, China University of Petroleum (East China), China

2 Energy & Geoscience Institute and Department of Chemical Engineering, The University of Utah, Salt Lake City, UT, USA

3 Sinopec Key Laboratory of Shale Oil/Gas Exploration and Production Technology, Beijing, China

4 Key Laboratory of Tectonics and Petroleum Resources of Ministry of Education (China University of Geosciences), Wuhan, China

5 Key laboratory of strategy evaluation for Shale gas, Ministry of Land and Resources, Beijing, China

6 PetroChina Exploration & Development Research Institute, Langfang, China

7 Faculty of Earth Resources, China University of Geosciences, Wuhan, China

8 School of Geoscience and Engineering, Southwest Petroleum University, Chengdu, China

9 School of Energy Resources, China University of Geosciences, Beijing, China

10 School of Ocean Sciences, China University of Geosciences, Beijing, China

References

[1] Van Wagoner J, Posamentier H, Mitchum R, Vail P, Sarg J, Loutit T, et al. An overview of the fundamentals of sequence stratigraphy and key definitions. In: Wilgus CK, Hastings BS, Kendall CGSC, Posamentier HW, Ross CA, Wagoner JCV, editors. Sea Level Changes: An Integrated Approach. 42. Tulsa, Oklahoma, USA: SEPM Special Publication; 1988. p. 39-45. DOI: 10.2110/pec.88.01.0039

[2] Vail PR, MitchumJr RM, Todd RG, Widmier JM, ThompsonIII S, Sangree JB, et al. Seismic stratigraphy and global changes in sea level. In: Payton CE, editor. Seismic Stratigraphy: Applications to Hydrocarbon Exploration. 26. Tulsa, Oklahoma: American Association of Petroleum Geologists; 1977. pp. 49-62

[3] Weimer P, Varnai P, Budhijanto FM, Acosta ZM, Martinez RE, Navarro AF, et al. Sequence stratigraphy of Pliocene and Pleistocene turbidite systems, northern Green Canyon and Ewing Bank (offshore Louisiana), northern Gulf of Mexico. AAPG Bulletin. 1998;**82**(5):918-960

[4] Posamentier HW, Allen GP. Siliciclastic Sequence Stratigraphy: Concepts and Applications. Tulsa, Oklahoma: Society for Sedimentary Geology; 1999

[5] Jiang S, Henriksen S, Wang H, Lu Y, Ren J, Cai D, et al. Sequence-stratigraphic architectures and sand-body distribution in Cenozoic rifted lacustrine basins, east China. AAPG Bulletin. 2013;**97**(9):1447-1475. DOI: 10.1306/03041312026

[6] Bohacs KM, Schwalbach JR. Sequence stratigraphy of fine-grained rocks with special reference to the Monterey formation. In: Schwalbach JR, editor. Sequence Stratigaphy in Fine-Grained Rocks: Examples from the Monterey Formation. Tulsa, Oklahoma: Society for Sedimentary Geology; 1992. pp. 7-19

[7] Wignall PB, Maynard JR. The sequence stratigraphy of transgressive black shales. Source Rocks in a Sequence Stratigraphic Framework. 1993;**37**:35-47

[8] Slatt RM, Rodriguez ND. Comparative sequence stratigraphy and organic geochemistry of gas shales: Commonality or coincidence? Journal of Natural Gas Science and Engineering. 2012;**8**:68-84. DOI: 10.1016/j.jngse.2012.01.008

[9] Hemmesch NT, Harris NB, Mnich CA, Selby D. A sequence-stratigraphic framework for the Upper Devonian Woodford Shale, Permian Basin, west Texas. AAPG Bulletin. 2014;**98**(1):23-47. DOI: 10.1306/05221312077

[10] Ver Straeten CA, Brett CE, Sageman BB. Mudrock sequence stratigraphy: a multi-proxy (sedimentological, paleobiological and geochemical) approach, Devonian Appalachian Basin. Palaeogeography, Palaeoclimatology, Palaeoecology. 2011;**304**(1):54-73. DOI: 10.1016/j.palaeo.2010.10.010

[11] Treadgold G, McLain B, Sinclair S. Eagle ford shale prospecting with 3D seismic data. SEG Technical Program Expanded Abstracts 2010: Society of Exploration Geophysicists; 2010. p. 2270-2273. DOI: 10.1190/1.3513302

[12] Baruch ET, Slatt RM, Marfurt KJ. Seismic stratigraphic analysis of the Barnett Shale and Ellenburger unconformity southwest of the core area of the Newark East field, Fort Worth Basin, Texas. In: J. Breyer, editor. Shale Reservoir: Giant Resources for the 21st Century. 97. Tusla, Oklahoma: AAPG Memoir; 2012. p. 403-418. DOI: 10.1306/13321483M97441

[13] Bowker KA, Grace T, editors. The downside of using GR to determine TOC content: An example from the Marcellus shale in SE West Virginia. In: Critical Assessment of Shale Resource Plays. AAPG Hedberg Research Conference; 2010; Austin, Texas.

[14] Algeo TJ, Schwark L, Hower JC. High-resolution geochemistry and sequence stratigraphy of the Hushpuckney Shale (Swope Formation, eastern Kansas): Implications for climato-environmental dynamics of the Late Pennsylvanian Midcontinent Seaway. Chemical Geology. 2004;**206**(3):259-288. DOI: 10.1016/j.chemgeo.2003.12.028

[15] Buckner N, Slatt RM, Coffey B, Davis RJ. Stratigraphy of the Woodford Shale from behind-outcrop drilling, logging, and coring. AAPG Search and Discovery Article. 2009; 50147

[16] Macquaker JH, Taylor KG, Gawthorpe RL. High-resolution facies analyses of mudstones: Implications for paleoenvironmental and sequence stratigraphic interpretations of offshore ancient mud-dominated successions. Journal of Sedimentary Research. 2007;**77**(4):324-339. DOI: 10.2110/jsr.2007.029

[17] Hammes U, Carr DL. Sequence stratigraphy, depositional environments, and production fairways of the Haynesville shale-gas play in East Texas. AAPG Search and Discovery Article. 2009;110084

[18] Hammes U, Hamlin S, Eastwood R. Facies characteristics, depositional environments, and petrophysical characteristics of the Haynesville and Bossier shale-gas plays of east Texas and northwest Louisiana. Houston Geological Society Bulletin. 2010;**52**:59-63

[19] Pyles D, Slatt R. Stratigraphic evolution of the Upper Cretaceous Lewis Shale, southern Wyoming: Applications to understanding shelf to base-of-slope changes in stratigraphic architecture of mud-dominated, progradational depositional systems. In: Nilsen TRS, Steffens G, Studlick J, editors. Atlas of Deepwater Outcrops, American Association of Petroleum Geologists Studies in Geology. 562009. pp. 585-489

[20] Slatt RM, Abousleiman Y. Merging sequence stratigraphy and geomechanics for unconventional gas shales. The Leading Edge. 2011;**30**(3):274-282. DOI: 10.1190/1.3567258

[21] Abouelresh MO, Slatt RM. Lithofacies and sequence stratigraphy of the Barnett Shale in east-central Fort Worth Basin, Texas. AAPG Bulletin. 2012;**96**(1):1-22. DOI: 10.1306/04261110116

[22] Slatt RM. Sequence stratigraphy, Geomechanics, microseismicity, and geochemistry relationships in unconventional resource Shales. Unconventional Resources Technology Conference; Denver, Colorado: Unconventional Resources Technology Conference (URTEC). 2014. DOI: 10.15530/urtec-2014-1934195

[23] Kohl D, Slingerland R, Arthur M, Bracht R, Engelder T. Sequence stratigraphy and depositional environments of the Shamokin (Union Springs) member, Marcellus formation, and associated strata in the middle Appalachian Basin. AAPG Bulletin. 2014;**98**(3):483-513. DOI: 10.1306/08231312124

[24] Chen L, Lu Y, Jiang S, Li J, Guo T, Luo C, et al. Sequence stratigraphy and its application in marine shale gas exploration: A case study of the lower Silurian Longmaxi formation in the Jiaoshiba shale gas field and its adjacent area in southeast Sichuan Basin, SW China. Journal of Natural Gas Science and Engineering. 2015;**27**:410-423. DOI: 10.1016/j.jngse.2015.09.016

[25] Jiang S, Xu Z, Feng Y, Zhang J, Cai D, Chen L, et al. Geologic characteristics of hydrocarbon-bearing marine, transitional and lacustrine shales in China. Journal of Asian Earth Sciences. 2016;**115**:404-418. DOI: 10.1016/j.jseaes.2015.10.016

[26] Hammes U, Frébourg G. Haynesville and Bossier mudrocks: A facies and sequence stratigraphic investigation, East Texas and Louisiana, USA. Marine and Petroleum Geology. 2012;**31**(1):8-26. DOI: 10.1016/j.marpetgeo.2011.10.001

[27] Van Wagoner JC, Mitchum R, Campion K, Rahmanian V. Siliciclastic sequence stratigraphy in well logs, cores, and outcrops: Concepts for high-resolution correlation of time and facies. In: Van Wagoner JC, Mitchum R, Campion K, Rahmanian V, editors. AAPG Methods in Exploration Series. 7. Tulsa, Oklahoma: AAPG; 1990. pp. 1-63. DOI: 0-89181-657-7

[28] Curtis JB. Fractured shale-gas systems. AAPG Bulletin. 2002;**86**(11):1921-1938. DOI: 10.1306/61EEDDBE-173E-11D7-8645000102C1865D

[29] Bustin M, editor. Geology report: where are the high-potential regions expected to be in Canada and the U.S. ? Capturing opportunities in Canadian shale gas. The Canadian Institute's 2nd Annual Shale Gas Conference; 2006; Calgary.

[30] Passey QR, Bohacs K, Esch WL, Klimentidis R, Sinha S, editors. From oil-prone source rock to gas-producing shale reservoir-geologic and petrophysical characterization of unconventional shale gas reservoirs. CPS/SPE International Oil and Gas Conference and Exhibition in China. Beijing, China: Society of Petroleum Engineers; 2010. DOI: 10.2118/131350-MS

[31] Jarvie DM. Shale resource systems for oil and gas: Part 1—Shale-gas resource systems. In: J. Breyer, editor. Shale Reservoirs—Giant Resources for the 21st Century. 97. AAPG Memoir; 2012. p. 69-87. DOI: 10.1306/13321446M973489

[32] EIA. EIA/ARI World Shale Gas and Shale Oil Resource Assessment Technically Recoverable Shale Gas and Shale Oil Resources: An Assessment of 137 Shale Formations in 41 Countries outside the United States. Washington DC: US Energy Information Administration; 2013

[33] Sonnenberg SA. The Niobrara petroleum system: A new resource play in the Rocky Mountain region. In: Estes-Jackson JE, Anderson DS, editors. Revisiting and revitalizing the Niobrara in the Central Rockies: Rocky Mountain Association of Geologists Guidebook. 2011. pp. 13-32

[34] Singh P. Lithofacies and Sequence Stratigraphic Framework of the Barnett Shale, Northeast Texas [Thesis]. The University of Oklahoma; 2008

[35] Lash GG, Engelder T. Thickness trends and sequence stratigraphy of the Middle Devonian Marcellus Formation, Appalachian Basin: Implications for Acadian foreland basin evolution. AAPG Bulletin. 2011;**95**(1):61-103. DOI: 10.1306/06301009150

[36] Li S, Cheng S, Yang S, Huang Q, Xie X, Jiao Y, et al. Sequence Stratigraphy and Depositional System Analysis of the Northeastern Ordos Basin. Beijing: Geological Publishing House; 1992

[37] Sitian L, Shigong Y, Jerzykiewicz T. Upper Triassic-Jurassic foreland sequences of the Ordos basin in China. In: Dorobek SL, Ross GM, editors. Stratigraphic Evolution of Foreland Basins. 52: SEPM Special Publication. 1995. pp. 233-241. DOI: 10.2110/pec.95.52.0233

Chapter 3

Sequence Stratigraphy of Fluvial Facies: A New Type Representative from Wenliu Area, Bohai Bay Basin, China

Jingzhe Li and Jinliang Zhang

Additional information is available at the end of the chapter

http://dx.doi.org/10.5772/intechopen.71149

Abstract

The application of sequence stratigraphy to the fluvial portion of sedimentary basin fills is most challenging, especially where the fluvial deposits under analysis are isolated or far away from coeval shorelines and marine influences. Taking the Wenliu Area as an example, this chapter aims at addressing researches about another type of river. High-resolution stratigraphic analysis of the lower second member of the Shahejie Formation of the W79 Block of Bohai Bay Basin (China) has revealed that the study area, previously interpreted as a shallow water delta system, actually originated in a subaerial setting with a distributional pattern. The base level fluctuations are mainly controlled by the regional tectonic setting. Active subsidence stages tend to make base level rising semi-cycles, while relative stable stages tend to make base level falling semi-cycles.

Keywords: fluvial, sequence stratigraphy, Bohai Bay Basin

1. Introduction

At the end of 1980s, Vail proposed the theory of sequence stratigraphy. At present, this theory has become a powerful new method to study sedimentary strata. Since Posamentier and Vail et al. in 1988 established the first non-marine (alluvial) sequence stratigraphy model [1], people began to pay attention to the difference between allogenetic cycles and authigenic cycles in fluvial depositional system. The eustasy, the tectonism, the climate, and other factors beyond the river power system are defined as allogenetic factors. The factor of the system itself caused by sediment cycles is called authigenic factor, such as the lateral migration of the channels. People can explain the interaction of the allogenetic controls (accommodation space and sediment supply) which have controlled the formation the fluvial sequence. Many authors have put forward their respective models [1–15]. But till now, there is still not a widely accepted model as a passive continental margin model of sequence stratigraphy.

Fluvial deposits are among the best understandable depositional systems. However, the associated sequence stratigraphy application in fluvial systems is one of the most challenging topic, especially when the river sediments are isolated or away from the shorelines and the sea, such as the overfilled continental sedimentary basins which have only the continental strata record preserved, or in the case that the data availability is limited to the continental basin. In the latter case, all efforts should be made to expand the range of observation scale, if possible, to study the relationships between the rivers and the age of the coastal and marine systems. However, the modern stratigraphy is enough in order to provide a genetic stratigraphic analysis of the stratigraphic record, from the scale of a single depositional system to the scale of the whole basin [16].

All siliciclastic detritus is transported by fluvial system from some point. Debris is eroded from mountain source area and is carried into basins, forming alluvial fans, coastal plains, deltas, fan deltas, as well as fluvial deposition itself. Recognizing fluvial depositional settings and reconstructing local fluvial styles are the main works of a sedimentologist. Three conventional examples [16] of fluvial styles are shown in **Figure 1**, while there may be some other types beside the conventional ones [19].

Taking the Wenliu Area as an example, this chapter aims at addressing researches about another type of river.

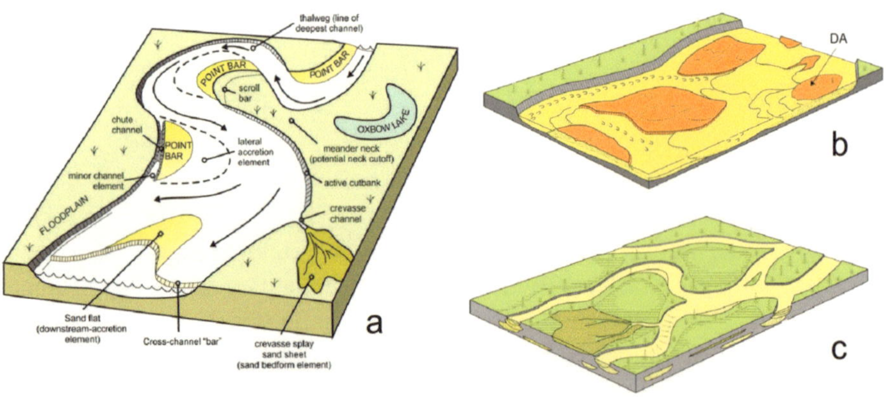

Figure 1. (a) A model for a meandering river, based on the South Saskatchewan River, Canada [17]; (b) a model for a sandy-braided river, based on the South Saskatchewan River, Canada [17]; (c) a model for an anastomosed river in a humid environment. Based on the work of Smith and Smith [18] in Banff National Park Canada.

2. Distributive fluvial deposition

The new river type we are going to propose is a kind of river that resembles distributive fluvial system (DFS) very much. DFS was firstly raised by Weissmann et al. [20]. Weissmann et al. have discussed that many aggradational depositional systems are dominated by distributive landscapes in both subaerial and subaqueous settings. The term "distributary fluvial systems" (DFS) addressed a key question in the earth sciences, i.e., that the most of the current fluvial facies models have a limited relevance in the interpretation of the ancient deposits.

Weismann et al. [20] also pointed out that many of the DFSs have been impacted by significant Quaternary climatic fluctuations and the deposits will represent the Quaternary sedimentary features of the history. From the regional to the local scale, the distributive fluvial systems display characteristics including: (1) deposition in the alluvial system becomes diving into the basin, (2) the radiation pattern of channels from a vertex, (3) a broadly fan-shaped deposit that is convex upward across the DFS and concave upward down-fan, and (4) an intersection point above which the alluvial system is held in an incised valley and below which it spreads out across an active depositional lobe (if the river is presently incised into its DFS). The observed DFS geometry and graphic style in these different basins are very different. In the open DFSs, the river is not confined in a valley, nor is able to transfer on the surface of DFS. The tearing process is dominated by node tearing near the DFS vertex or intersection. These rivers seem to have significant differences from the valley, because the floodplain material can be easily deposited and saved in the open fan. The DFS rivers observed in many sedimentary basins are now wholly or partially cut into the DFS. The incision takes two main forms: (1) incision of the proximal DFS controlled by sediment supply and discharge control due to climate change [21, 22, 23] or tectonic tilting [21, 22], or (2) incision driven by basic level decline (whether is the ocean, lakes, or by river capture) [24].

3. Sequence stratigraphic model of fluvial facies

Fluvial sedimentation is the result of several allogenic factors, including sea level, environmental energy flux, source area tectonic movement, and basin subsidence [3, 4] (**Figure 2**). The relative importance of these factors is hard to determine, although related stratigraphic criteria can be derived from field studies and from experimental work [25]. The tectonism, the relative sea level fluctuations, and the climatic controls may be interpreted from the changes through geologic time in the directions of tectonic tilting during the deposition and from the variations of the landscape gradients. They can also be interpreted from the variations in the depths of burial, as inferred from the analyses of paleocurrent directions, architectural elements, fluvial styles, and late diagenetic clay minerals, coupled with isotopic geochemistry and petrographic studies of framework and early diagenetic constituents [26–28].

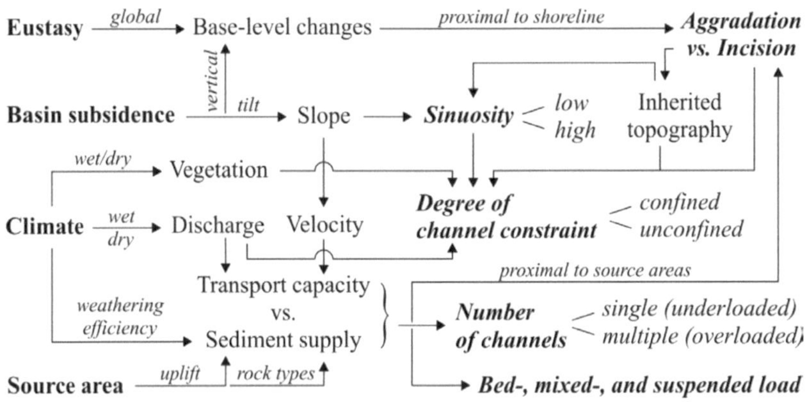

Figure 2. Allogenic controls on fluvial sedimentation [16].

The base level control on fluvial cyclicity represents the bulk of the first-generation sequence stratigraphic models, which assume a direct correlation between rising and falling base level on one hand, and fluvial aggradation and downcutting on the other hand [1, 29, 30]. The predictable relationship between fluvial processes and base level changes reflects a most likely scenario, but exceptions do exist [16]. This relationship is valid for the downstream reaches of fluvial systems, where rivers respond to "downstream controls" (i.e., interplay of sea level changes, basin subsidence, and fluctuations in environmental energy flux induced by climate change). In such settings, which may be characterized by either low or high accommodation space in the Leckie and Boyd's [31] scheme of fluvial stratigraphy, the fluvial deposits may be integrated within the standard lowstand, transgressive, and highstand systems tracts.

Wright and Marriott [2] believe that lowstand system tracts are mainly composed of amalgamated channel deposits. The thick transgressive systems tracts are characterized by the deposits of floodplains that are mainly wrapped with isolated channel sand bodies. The upward highstand systems tracts are characterized by higher density sandstones and paleosol deposits, representing good stratigraphic markers. Shanley and McCabe [4] considered that fluvial strata can be traced to marine strata from the same period. Laterally amalgamated fluvial sheet sandstone can overlap the unconformities. In the upper alluvial parts, amalgamated channel deposits turn into relative isolated and fine-grained sediments interbedded with meandering channel deposits. This sedimentary feature shows the tidal influence from the ocean. Catuneanu [16] made a summary of these two representative models (**Figure 3**). Thus, the fluvial sequence boundaries are placed at the bottom of base level cycles. Incised valleys form when the base level falls, while lowstand systems tracts occur at the beginning of base level rise [32].

Holbrook et al. [11] introduced the useful buttresses and buffer concepts to explain longitudinal changes in fluvial facies and building upstream coastline. One is some fixed point of control on the river equilibrium profile in the ocean basin. It is a base level (sea level) in the main river inland basin. The reaction buffer space with and below the current gradient profile

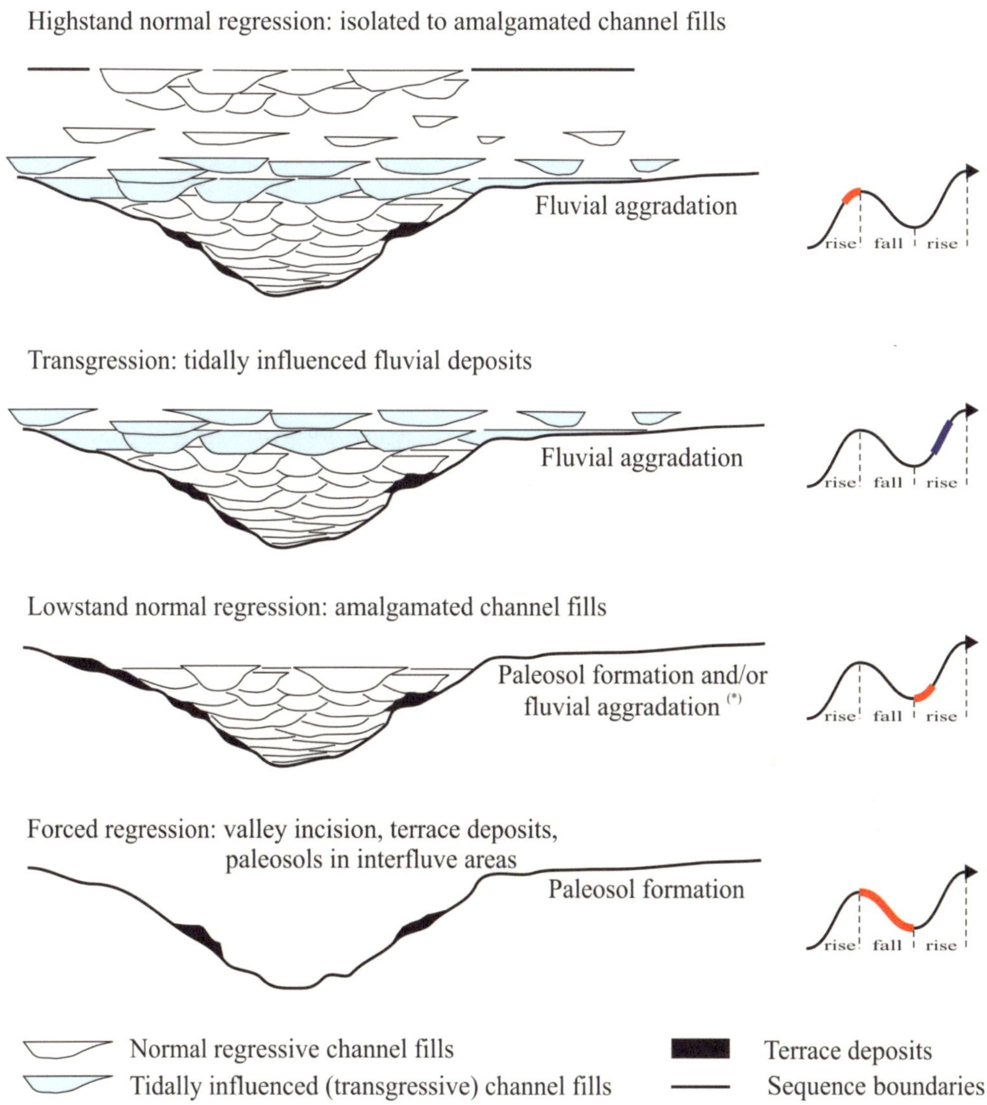

Figure 3. Stratigraphic model of fluvial facies modified from [16].

as the representative of the file may not exist an upstream control, such as tectonic or climatic change, the influence of river flow, and sediment. The tectonic uplift may increase the gradient distribution and the upper buffer zone. The decreasing of the butresses, such as in the fall of sea level, may lead to the river system incision, but if the sea level fall blow newly exposed continental shelf as a similar slope, the river profile, there may be little change in the style of the river. In this case, the river system will change into to a new dynamic balance due to new water and sediment flux rate. In the accommodation between the upper and lower areas of the buffer zone, a representative (potential) of the river system to save space is available.

The ratio between channel and floodplain architectural elements during stages of positive accommodation depends on the rates of base level rise. Rapid base level rise leads to increased floodplain aggradation, which results in overall finer-grained successions. Slowly rising base level creates little accommodation available in overbank areas. At the same time, the channel stack in reducing accommodation time may be accompanied by frequent avulsion, which contributes to the spread of excessive lateral sediment [33]. Channel amalgamation under conditions of low accommodation is usually the case with the lowstand and late highstand systems tracts. As the late highstand amalgamated channel fills have a low preservation potential due to the subsequent erosion associated with the subaerial unconformity, the fluvial portion of the depositional sequence commonly displays a fining-upward profile (**Figure 3**). These general principles of fluvial stratigraphy, which relate the stacking patterns of fluvial architectural elements to changes in base level and available accommodation, have also been documented in the case of fan delta systems, which are governed by similar process/response relationships between fluvial processes within alluvial fans and the base level fluctuations of the standing bodies of water into which they prograde [34].

4. Methodology

4.1. Sequence stratigraphic analysis

Geophysical borehole logs represent various rock properties, which can be used for stratigraphic interpretation. The most common type of log, often used for local stratigraphic correlation, is summarized in **Figure 4**. As technology improves, some new types of well logs are being developed. For example, the new micro-resistivity logs combine the methods of conventional resistivity and dipmeter measurements to produce high-resolution images that simulate the sedimentological details of an actual core. This "virtual" core allowed in the details of mm scale visualization, including sediment bedding, cross-bedding, and biological noises. Well logs have both merits and disadvantages compared with outcrops. Geophysical logging in the outcrop of the advantage is that they provide continuous information from the inheritance of relatively thick, often in a range of kilometers. This type of configuration file (log curve) allows one to see the trend in different scales, from single element sedimentary deposition system in the size of the whole basin, to fill. For this reason, data provided by well logs may be considered more complete relative to the discontinuous information that may be extracted from the study of outcrops. Therefore, comparative study of underground relation and formation can usually scale far greater than from the outcrop research. On the other hand, nothing can replace the study of the actual rocks; hence, the wealth of details that can be obtained from outcrop facies analysis cannot be matched by well-log analysis, no matter how closely spaced the boreholes may be [35].

The surface of log interpretation is largely speculative, under the condition of practical rock data. The core data provide the most clear "ground truth information" [36]. Therefore, the

Log	Property measured	Unit	Geological uses
Spontaneous potential	Natural electric potential (compared to drilling mud)	Millivolts	Lithology (in some cases), correlation, curve shape analysis, identification of porous zones
Resistivity	Resistance to electric current flow	Ohm-metres	Identification of coals, bentonites, fluid evaluation
Gamma-ray	Natural radioactivity-related to K, Th, U	API units	Lithology (shaliness), correlation, curve shape analysis
Sonic	Velocity of compressional sound wave	Microseconds/metre	Identification of porous zones, coal, tightly cemented zones.
Caliper	Size of hole	Centimetres	Evaluate hole conditions and reliability of other logs
Neutron	Concentrations of hydrogen (water and hydrocarbons) in pores	Percent porosity	Identification of porous zones, cross-plots with sonic, density logs for empirical separation of lithologies
Density	Bulk density (electron density) includes pore fluid in measurement	Kilograms per cubic metre (gm/cm^3)	Identification of some lithologies such as anhydrite, halite, non-porous carbonates
Dipmeter	Orientation of dipping surfaces by resistivity changes	Degrees (and direction)	Structural analysis, stratigraphic analysis

Figure 4. Types of well logs, properties they measure, and their use for geological interpretations (after [17, 18, 35]).

geophysical data including logging and seismic can only provide indirect information on the solid and fluid phases in the underground that must be calibrated and the interpretation accuracy of verification to verify the geological data of rock [16]. The integration of all available data sets (e.g., outcrop, core, logging, and cuttings, therefore, the seismic) is the best way to correctly identify the stratigraphic contact.

The seismic data can also be used to produce seismic stratigraphic interpretation in the two kinds of materials at the interface of different acoustic characteristics. This can be described by acoustic impedance and is given by the product of density and velocity. The larger is the difference in acoustic impedance between two lithologies, and the stronger is the reflection. Various subsurface layers can be recorded by seismic traces through geophones that receive the reflections. The impedance contrasts control the nature of each wavelet. Such seismograms are time seismic sections. The seismic section generally follows the empirical Faust formula of seismic wave velocity, depth, and age proof [37]:

$$V = 46.5(ZT)^{1/6}\text{m/s}, \tag{1}$$

V is seismic velocity, Z is buried depth, and T is geological age.

Continuous seismic reflection is a "time line" hypothesis; in the basin, it is of great significance to analyze and research the sequence stratigraphy. Seismic records are also complicated by multiples and diffractions. Multiples may be strong enough to obscure deep reflections, but they are now relatively easy to remove by processing. The diffraction from the steep inclined surface is the scripture, such as fault, channel profit, and the erosion of relief unconformity. They are useful indicators of other components showing a steep reflection event but may be confused by the migration. The seismic record can be considerably improved, and interpretation facilitated by the use of two special techniques, the construction of synthetic seismograms, and the use of vertical seismic profiling (VSP) [38].

Synthetic seismograms are generated by the conversion of sonic and density data reflection coefficients. VSP is the recording and analysis of seismic signals received from a geophone lowered downhole. The well is reduced to be recorded on the surface of the signal, when they return to the surface of the detector to incrementally shift. Still, the shortest wavelength signal to each half the length of the T detector was analyzed. Each location has a new record. The synthesis and VSP data is very valuable for calibration of seismic records. For example, they enable a detailed record of seismic velocities to produce depth-corrected cross-sections. However, the VSP technique is far better for various reasons. It can, for example, can be used to calculate the total depth of expansion (TD) of synthetic seismic records, and not. Synthesis methods rely on logging data A, only reflect the conditions close to the hole. In the case of poor hole conditions or seismic reflectors having a very small horizontal extent (less than one Fresnel zone), the synthetic method may not provide a record typical of the area.

In this chapter, the method of stratigraphic analysis we used is an integrated well-seismic correlation. The Wenliu Area possesses abundant available materials, including a 3D seismic database (covering an area of about 35 km^2), more than 340 drilling wells and logging data (including core data of over 330 m from 8 cored wells of the area), and plentiful analysis test data. Thus, it is possible to launch a detailed sequence stratigraphic study.

With the 3D seismic database, regional reflection surfaces were traced to the target area. With the correlation of synthetic seismogram and VSP logging data, the relationship between seismic travel time and well depth was established. More than 10 seismic well sections were interpreted and main structures (especially faults) were identified. A more precise analysis was conducted with logging data beyond this. Wavelet transformation and Maximum Entropy Spectral (MEM) analysis with gamma-ray data were also used to enhance the accuracy of the sequence identification.

4.2. Facies mapping

Logging data are used for stratigraphic and lithologic interpretation, while they can also be used directly in facies mapping. Lithologic information may yield different combinations of logs in the CRO field reconnaissance. These relationships can be converted into a computer algorithm and the log data are digitized and stored in the data bank, a powerful automatic mapping technology. Digital log data can also be displayed and manipulated using interactive computer graphics routines, a facility which can easily compare formation related purposes. Well service company invested a lot of money on design and marketing automation processing and display used in basin analysis and petroleum development techniques, but these techniques suffer from the limited resolution of the physical location is very special. The techniques cannot be used without much initial careful calibration to local petrographic and groundwater conditions.

In addition to logging method as the foundation, extensive additional technique has been developed for underground rock physics for many years, the core and sample data check. These methods have several goals: such as stratigraphic correlation, provenance, reconstructions of paleogeography, regional stratigraphic trends, and tectonic history. Some of these techniques provide a numerical study of the age information. Others are useful or relevant local or regional origin but do not necessarily provide such age information. The grain size, grain shape, and the debris of a stratigraphic unit depend on the initial nature of the clastic source. Yet, after transportation, deposition, and burial process, the debris may experience many metamorphic processes, so that the features of the original source blurred. The analysis of basin detrital composition depends on these main factors: (1) source area geological structure; (2) source area of the climate and terrain; (3) transport process of debris diffusion and mixing mode brings; (4) chemical and mechanical wear, winnowing, transport, and deposition; and (5) diagenetic changes during burial of deposits.

The seismic data can also be used for basin mapping. The seismic reflector, the amplitude, and the continuity of seismic facies are the three elements, becoming more and more important to improve the processing and visualization of seismic sections. The concept of seismic facies is the most effective application in the main data including 2D cross-section, but the stratigraphic and sedimentological interpretation is not easy in two-dimensional settings. This problem is not too serious when 3-D data is available, because the nature of the whole volume of 3-D data could be a visual system to provide guidance for the real world. The calibration between the seismic attributes and the lithofacies is specific for each basin to a considerable extent.

The channel environment 3D interpretation of sedimentary system is the most spectacular one, especially in the alluvial and in the submarine fan depositional settings, in the channel-levee complexes and crevasse splays, and finally in the coastal plains [39].

In this chapter, the methods to the facies analysis and mapping we used are traditional mapping approach. One of the first step in the facies analysis of a clastic reservoir is the description and interpretation of available conventional cores. Core description was based on 10 cores taken from 7 cored wells of the target interval of the Wenliu Area. The color, the sedimentary structures, and the grain size of deposits were analyzed (the grain size analysis was based on a LS130 Coulter laser micro-granulometer). Sedimentary source analysis was also conducted with heavy mineral combination analysis, which calculates the ratio of stable transparent heavy minerals (mainly zircon and tourmaline) to the entire transparent heavy minerals (zircon, tourmaline, garnet, barite, epidote, etc.). Sedimentary facies were interpreted based on the analysis above. Finally, based on the detailed stratigraphic framework and counted sand body data, the lithofacies paleogeographic characteristics were analyzed, and along with the interpreted types of facies, a depositional model was established and the possible distribution of sedimentary facies was predicted both laterally and vertically.

5. Geological background of the Wenliu Area

The Bohai Bay Basin is located in the eastern China continental margin. It is a composite sedimentary basin formed in the Mesozoic and Cenozoic periods and the Proterozoic to the

Paleozoic period, covering an area of 195,000 square kilometers in the Bohai Sea and the coastal areas. Various reservoir systems and reservoir rock types are well developed in the basin, and the Eocene Shahejie Formation is the major hydrocarbon bearing layer series (**Figure 5a**) [40]. Some new studies show that the formation and evolution of the Bohai Bay Basin was induced by both the thermal power of bottom of the lithosphere and plate boundary activity due to regional stress field. The dynamic process of the Cenozoic basin in the Bohai Bay Basin is related to the thermal activity of the lithosphere in eastern China. It is also related to the regional stress field controlled by the tectonic activity at the plate boundaries and belongs to the superimposition of many regional geological processes [41, 42]. The Bohai Bay Basin is a Meso-Cenozoic sedimentary basin superimposed on the Paleozoic basement. During previous geological times the basin was part of the North China craton. The upper Proterozoic and the lower Paleozoic sequences are constituted of marine sediments, including the upper Proterozoic strata of the Jixian system, the Changcheng system and Qingbaikou system, which are mainly composed of micritic limestones, quartz sandstones, and shales. The sedimentary cover is not only distributed in the lower part of the basin, but widely distributed around the basin. The lower Paleozoic sequence is made up of Cambrian and Ordovician carbonates, with shales, siltstones, and sandstones. The upper Paleozoic sequence is made up of Carboniferous and Permian strata, with paralic and continental clastic rocks and some coal seams. The Mesozoic strata range in age from the Triassic to the Cretaceous. They are composed of tuffaceous sandstones, conglomerates, mudstones, and coals (**Figure 6**).

The Dongpu Depression, located in the south margin of Bohai Bay Basin, is an oil-rich sedimentary depression. It is a 16-km long and 70-km wide NNE fault depression that covers an area of approximately 5300 km^2 (**Figure 5a**). The Dongpu Depression formed from the Huabei Movement in the late Paleogene. The evolution of the depression reveals an evident tectonic component and shows periodic features [43]. It is bounded by the

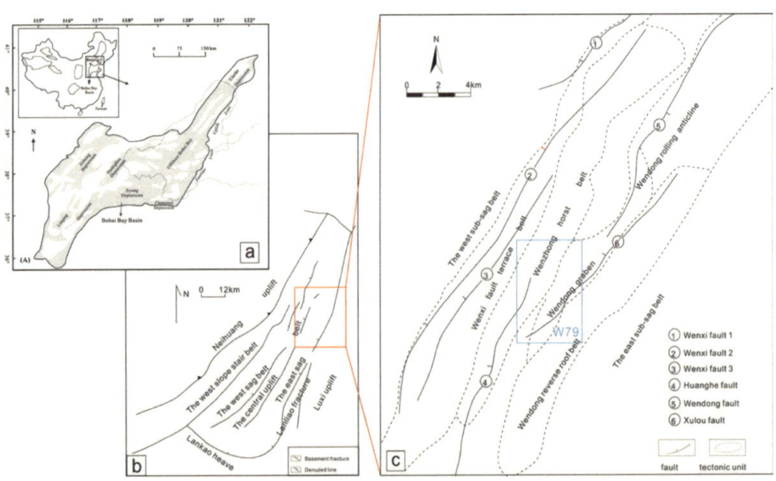

Figure 5. (a) Location of the Bohai Bay Basin (after [40]). (b) Location of the Dongpu Sag. (c) Location of Wenliu Area showing the main structures (after [19]).

http://dx.doi.org/10.5772/intechopen.71149

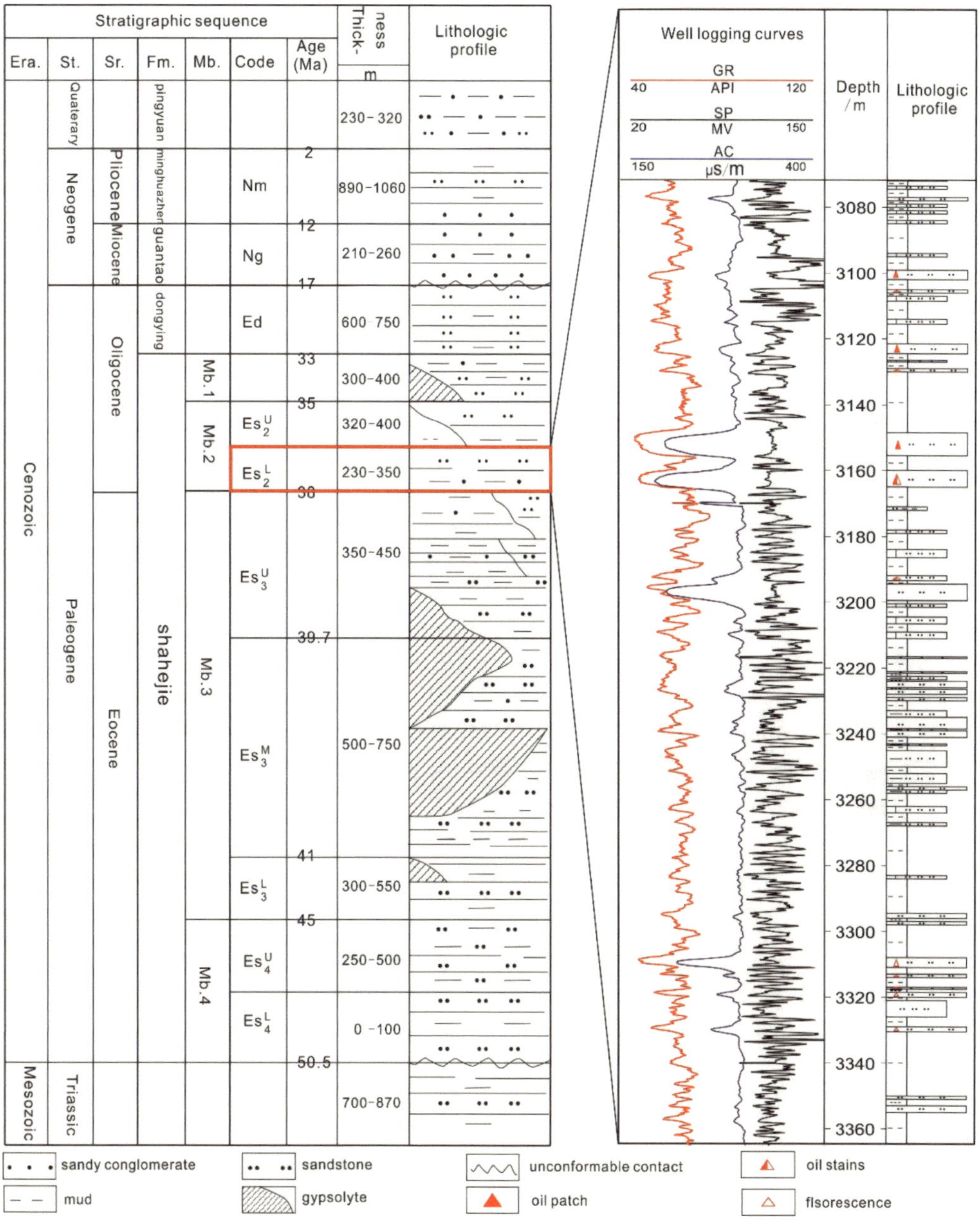

Figure 6. Stratigraphy in Dongpu Depression, Bohai Bay Basin (left), enhancing the lower second member of Shahejie FM. In Well W167 of Wenliu Area (right). St, system; Sr, series; Fm, formation; Mb, member (after [19]).

Luxi Uplift on the east, the Lankao Uplift on the south, and the Neihuang Uplift on the west. Its north margin is the Maling Fault, which is also the south margin of the Linqing Depression. The Lanling, Changyuan, and Huanghe basement rift faults controlled the early evolution of the Dongpu Depression and changed it from a broom fault depression to a double-break fault depression. During the early Paleogene, the Tan-Lu Faults changed from transpression to dextral lateral strike-slip faults, altering the regional stress field from a shear-extrusion field to a shear-tensile field. Differential subsidence also occurred during this period. The regional shear-tensile effects caused the Central Uplift rise, then induced gravitational sliding and salt rock intrusion, and finally molded the present appearance of the Dongpu Depression [44].

The Wenliu Oilfield, located in Henan Province, is on the northern part of the central uplift of the Dongpu Depression. Bounded by two large sag belts (i.e., the West and the East Sub-sag Belt), the Wenliu Oilfield is a complicated fault zone that is characterized by horst and graben structures [44]. It is also of the NNE direction and is approximately 20 km long and 16 km wide, covering an area of more than 2000 km^2 (**Figure 5b**).

The W79 Block, one of the most important blocks of the Wenliu Area, has already entered the middle to late stage of the developing period. The Paleogene lower second member of the Shahejie Formation (Es_2^L, lowermost Oligocene succession, ranging from approximately 2700 m to 3000 m) of this block, originally described as a shallow water delta system deposit, is the main oil-bearing interval of this area (the oil-bearing area is nearly 13.4 km^2) and has plentiful remaining petroleum resources. However, because of its complicated sedimentary features, many studies [41, 44–47] concerning sequence stratigraphy and sedimentary facies models are still under scientific debate. Hence, this paper aims to provide an analysis of the sequence stratigraphic and sedimentary characteristics of the Es_2^L, Paleogene Shahejie Formation of the W79 Block based on abundant well logging data, 3D seismic data, and core data under the principles of high-resolution sequence stratigraphy and sedimentology (**Figure 5c**).

6. Sedimentary model of Wenliu Area

The reservoir properties depend on tectonism, eustasy, sediment flux, biological process, climate, and other allogenic and authigenic factors. At the basin scale, these factors mainly determine the formation of systems tracts [48], while at smaller scales, they influence colors, shapes, structures, and internal architectures of the sedimentary bodies. It is at these smaller scales that reservoir characterization (especially lithological and lithofacies analysis) becomes significant.

According to core observation and description as well as well logging analysis, this chapter concludes that the lithology of the target area is characterized by mudstones, sandy mudstones, muddy siltstones, shaley siltstones, and sandstones. The main color of the sandstones is light brown, whereas the main color of the mudstones is red to purple. Statistics shows that mudstone is the dominant lithology of the target interval (more than 63%), with the other components showing relatively smaller percentages (siltstone 14.5%, muddy siltstone 5%, limy siltstone 5%, and sandstone 12.5%). Shapes of grain size cumulative curves plotted on log-probability paper could reveal environmental characteristics. The plots of grain size as cumulative curves of the target interval show that suspension population occupies a large proportion (generally larger than 30%). Saltation population is dominant and its segment is relatively steep. The truncation points between these two populations are usually bigger than 3 (φ value). Traction population is usually missing within the target interval (**Figure 7**).

The ratio of stable transparent heavy mineral assemblage (mainly zircon and tourmaline) to the entire transparent heavy mineral amount (zircon, tourmaline, garnet, barite, epidote, etc.) indicates that the sourcelands are on NNW direction.

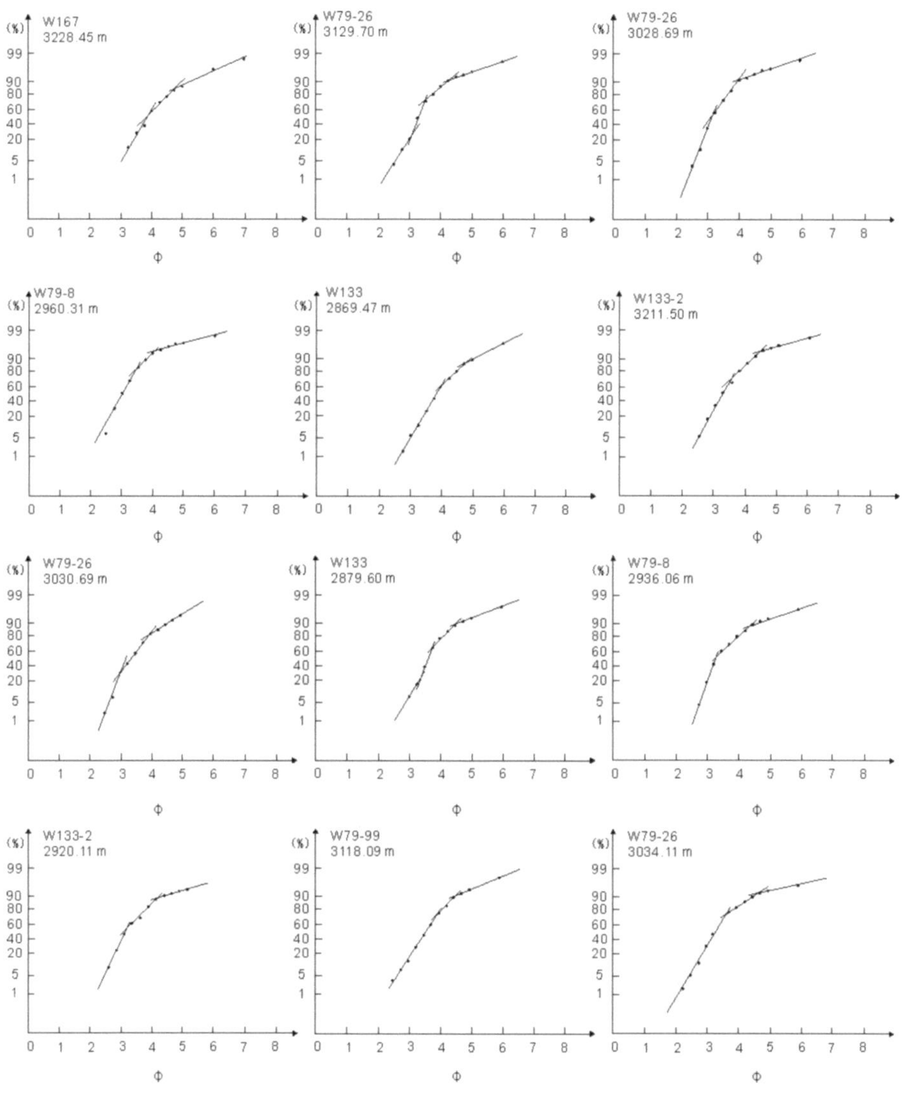

Figure 7. Grain size cumulative probability plots of target interval in Wenliu Oilfield (after [19]).

6.1. Lithofacies characteristics

The lithofacies characteristic analysis is also very important for depositional environment analysis [49–52]. The lithofacies recognition and categorization is based on lithology, grain size, physical and biogenic sedimentary structures, and stratification from core observation and description. Seventeen lithofacies have been recognized within the study interval, including from fine-grained sandstones of parallel bedding to mudstones of

Figure 8. Typical lithofacies of the W79 block for the target interval: (a) parallel bedding sandstone; (b) trough cross-bedding sandstone; (c) scouring structured sandstone with mud gravels; (d) ripple cross-bedding siltstone; (e) wavy cross-bedding siltstone; (f) parallel bedding siltstone; (g) ripple cross-bedding muddy siltstone; (h) flasher structured mudstone (after [19]).

lenticular structures (**Figure 8**). The codes, features and interpretation are summarized in **Table 1**.

According to the reddish-brown mud of subaerial oxidizing features and the grain size cumulative plots revealing typical channel deposit features (**Figure 7**), along with the specific

ID	Lithology	Structure and texture	Interpretation	Schematic representation
St	Sandstone	Trough cross-bedding	Channel lower part, curved bounding surfaces, originates by migration of 3-D bedforms, high energy	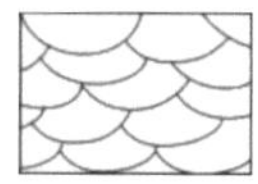
Sla	Sandstone	Low-angle cross bedding	Low-angle foreset lamina of single direction (generally lower than 10°), high-energy	
Sp	Sandstone	Parallel bedding	Parallel and aclinic, stable and high energy	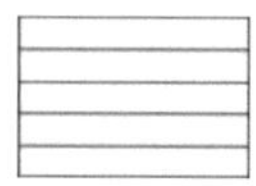
Ss	Sandstone	Scouring	Channel bottom, erosional bounding surface, usually mingled with mud gravels, high energy	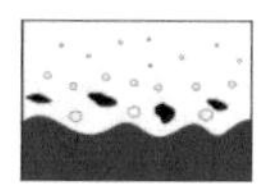
SSr	Siltstone	Ripple cross-bedding	Superimposition of one ripple on another as the ripples migrate, abundant sediment supply (especially suspension population)	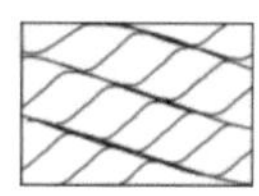
SSla	Siltstone	low-angle cross bedding	Low-angle foreset lamina of single direction (generally lower than 10°)	
SSp	Siltstone	Parallel bedding	Parallel and aclinic, result of suspension settling of fine-size sediment	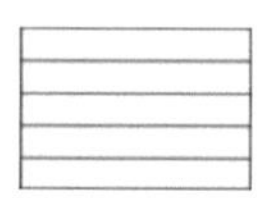
SSt	Siltstone	Trough cross bedding	Curved bounding surfaces, originates by migration of 3-D bedforms	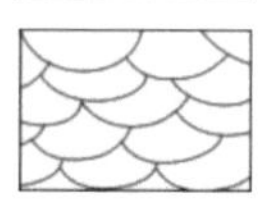
MSSr	Muddy siltstone	Ripple cross bedding	Superimposition of one ripple on another as the ripples migrate, abundant suspension sediment supply	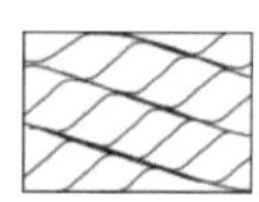
MSSw	Muddy siltstone	Wavy bedding	Wavy lamina, parallel to the bounding surface as a whole, abundant sediment supply of clay and silt	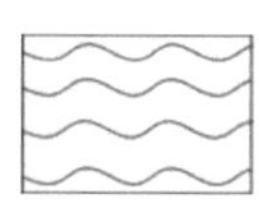

ID	Lithology	Structure and texture	Interpretation	Schematic representation
MSSp	Muddy siltstone	Parallel bedding	Parallel and aclinic, result of suspension settling of fine-size sediment osr precipitation from solution, low energy	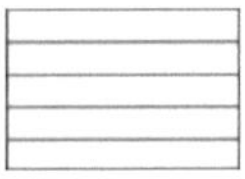
MSSla	Muddy siltstone	Low-angle cross bedding	Low-angle bounding surfaces, originates by migration of planar bedforms, low energy	
SMf	Silty mudstone	Flaser bedding	Fluctuating depositional conditions marked by periods of current activity, mud < sand	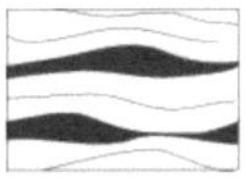
SMw	Silty mudstone	Wavy bedding	Wavy lamina, parallel to the bounding surface as a whole, abundant sediment supply of clay and silt	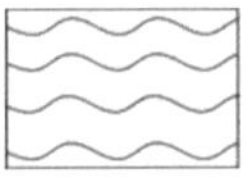
Mw	Mudstone	Wavy bedding	Wavy lamina, parallel to the bounding surface as a whole, abundant sediment supply of clay	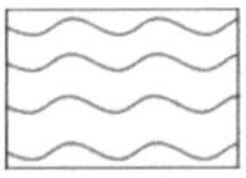
Mm	Mudstone	Massive bedding	Contain few or no visible internal lamina, overbank or abandoned channel deposits	
Ml	Mudstone	Lenticular bedding	Fluctuating depositional conditions marked by periods of current activity, mud > sand	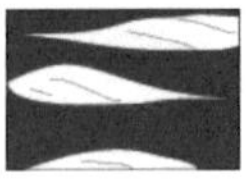

Table 1. Lithofacies of the target interval succession (after [19]).

associations of the lithofacies as reconstructed from the core descriptions, a meandering river fan is interpreted to be responsible for the sedimentation.

6.2. Meandering river fan

The target interval of the Dongpu Depression was formed under a rift contraction basin environment far from the source and with an abundant sediment supply. Single well core section shows that channel and overbank deposits are two main sequence types of the sedimentary strata. The channel deposits generally have complete sequences and are relatively isolated. The overbank deposits are mainly composed of reddish-brown shale that could indicate the exposed and oxidic environment (together with the lack of subaqueous paleontological fossils, it may be revealed as subaerial sedimentary environment) (**Figure 9**). Lithofacies paleogeographic mapping shows many meandering channels distributed along

http://dx.doi.org/10.5772/intechopen.71149

the basin margin like a large fan-shaped river system, which shows distributive characteristics quite different of those of a fan delta system (**Figure 10**).

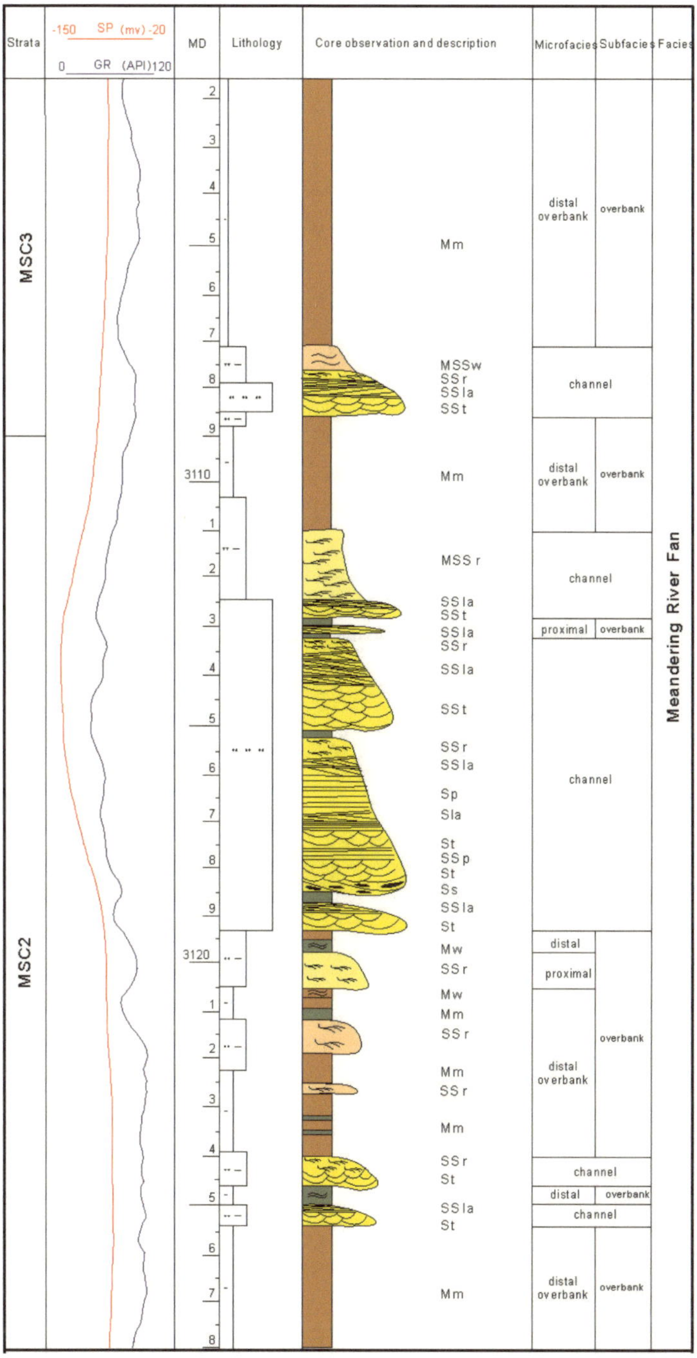

Figure 9. Single well facies (W133-12 Well) (after [19]).

This special depositional system is interpreted to be a meandering river fan, mainly composed of three subfacies (i.e., channel fill subfacies, overbank subfacies, and river flood lake subfacies) (**Figure 10**). In the study area, back-stepping and forward-stepping types both exist, revealing A/S > 1 and A/S < 1, respectively, and their logging curve (gamma ray) reflections are generally of toothed boxes or bell shapes (see **Table 2**).

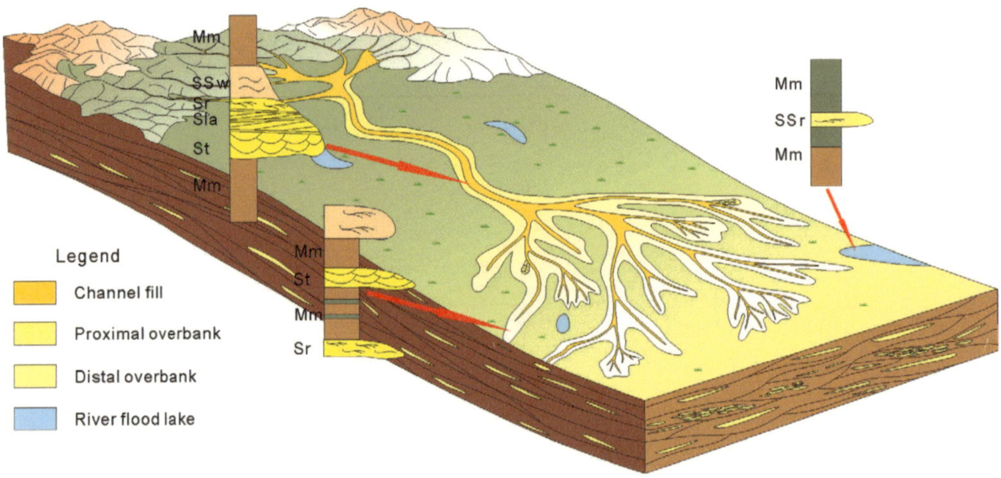

Figure 10. Sedimentary facies model of the target interval in Dongpu Depression (after [19]).

	Meandering river fan	Shallow water delta
Settings	Mainly subaerial	Subaqueous and subaerial
Paleo-climate	Humid; oxidizing environment (major part) to slightly reductive environment (some small parts)	Relatively humid; reductive environment to slightly oxidizing environment
Subaqueous paleontological fossils	Rare	Relatively abundant
Wave or tide forces	No	Yes
Sediment supply	Sufficient	Sufficient
Deposits between distributive/ distributary channels	Mud	Mud or sand sheet
Distance between distributive/ distributary channels	Relatively large	Small to medium
Sequence and cyclicity	Mainly positive cycles	Both positive and negative cycles
Subfacies	Channel fill subfacies, overbank subfacies and river flood lake subfacies	Delta plain subfacies, delta front subfacies, and pro-delta subfacies

Table 2. Comparison of meandering river fan and shallow water delta (after [19]).

The channel fill subfacies are mainly composed of sand-rich channel deposition. From bottom to top, Ss, St, Sla, and Sr constitute the main lithofacies and show a fining-upward cycle. The grain size is relatively coarse, and the sandbody connectivity is therefore high. The overbank subfacies contain two parts. The proximal overbank is next to the channel and is one of the most important reservoir types due to its good physical properties (i.e., after channel fill). The common lithofacies of the proximal overbank include Sla, SSla, and SSr. The distal overbank is distributed between channels and is mainly composed of purple to red mudstone. The common lithofacies include MSSw, SMw, and Mm. The river flood lake subfacies form near the distal overbank, recording stand water, and are mainly composed of grayish-green mudstone. The common lithofacies of this subfacies contain Mw, Mm, and Ml.

7. Sequence stratigraphy

7.1. Sequence stratigraphic surfaces

The identification of sequence stratigraphic surfaces and classification of depositional trends is the core component of high-resolution stratigraphic correlation and division. Integrated seismic and well analyses show that two major sequence stratigraphic surfaces collaboratively limit the target interval.

SB1, the top surface of the target interval, has strong and continuous reflection in the seismic section (**Figure 11a**). It is a regional parallel unconformity and is characterized by overlying thick and continuous shale of Es_2^U. Its features on the well logging curves are high GR (gamma ray), AC/DT (acoustic) and SP (spontaneous potential) values and a low R25 (2.5 m resistivity) value. Below this surface, tooth-shaped AC/DT and GR curves become the main characteristics, revealing a frequent sand mud inter-bedding (**Figure 11b1**).

SB2, the bottom surface of the target interval, also has strong and continuous reflection in the seismic section (**Figure 11a**). It is also a parallel unconformity characterized by a stable mudstone layer of the underlying Es_3^U. Its features on well logging curves are high GR, AC, SP, and R25 values (**Figure 6**). The most obvious characteristic is the in the SP curve, approximately 10 m above the surface, in which the curve has a v-shaped low value zone (**Figure 11b2**).

7.2. Facies associations under sequence stratigraphic frame

Based on core observations as well as integrated seismic and well analyses, more than 58 short-term base level cycles could be recognized in the target interval of the W79 Block. Genetic sequence stacking patterns can be used to define genetic sequence sets, showing base level rises and falls [1]. Thus, these short-term patterns could be combined into six middle-term base level cycles (from bottom to top—MSC1 and MSC6). In a similar way, given the regional setting, these six middle-term base level cycles could be further

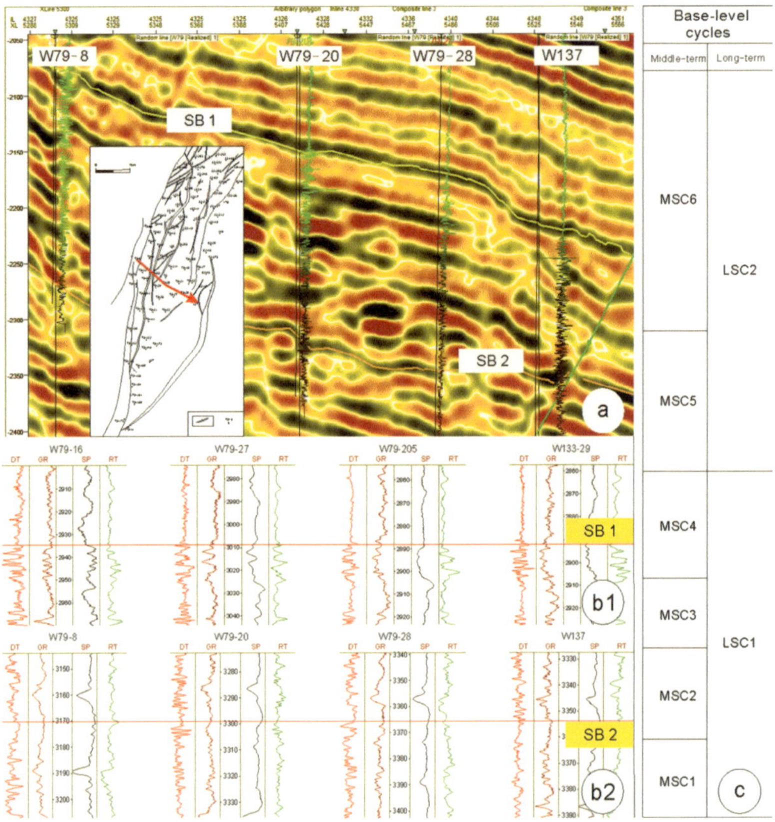

Figure 11. Identification of main sequence stratigraphic surfaces and stratigraphic division framework (after [19]).

combined into two long-term base level cycles (from bottom to top—LSC1 and LSC2) (**Figure 11c**). Under this high-resolution sequence stratigraphic frame, according to the practice of oilfield development, a 58-small-layer plan was accepted for paleogeographic mapping (**Figure 12**) and for other subsequent studies. Precise paleogeographic mapping shows that the vertical facies distribution and association have strong regulations on the long-term base level cyclic scale.

LSC1, the first long-term base level cycle, composed of MSC1 to MSC4, approximately 90 m to 120 m long, is the lower part of the target interval and is limited by the SB2 at the bottom. At the lower part of this long-term base level cycle, proximal overbank deposition and distal overbank deposition are most commonly found, whereas channel fill deposition could sporadically be found. In contrast, at the upper part of this long-term base level cycle, apart from proximal and distal overbank, channel fill deposition is also commonly found and the connectivity of the sandbody is relatively higher than that of the lower part of LSC1, which has no or isolated channel fill deposition. The ratio of channel fill to overbank increases from bottom to top, and the sedimentary stratigraphy shows an

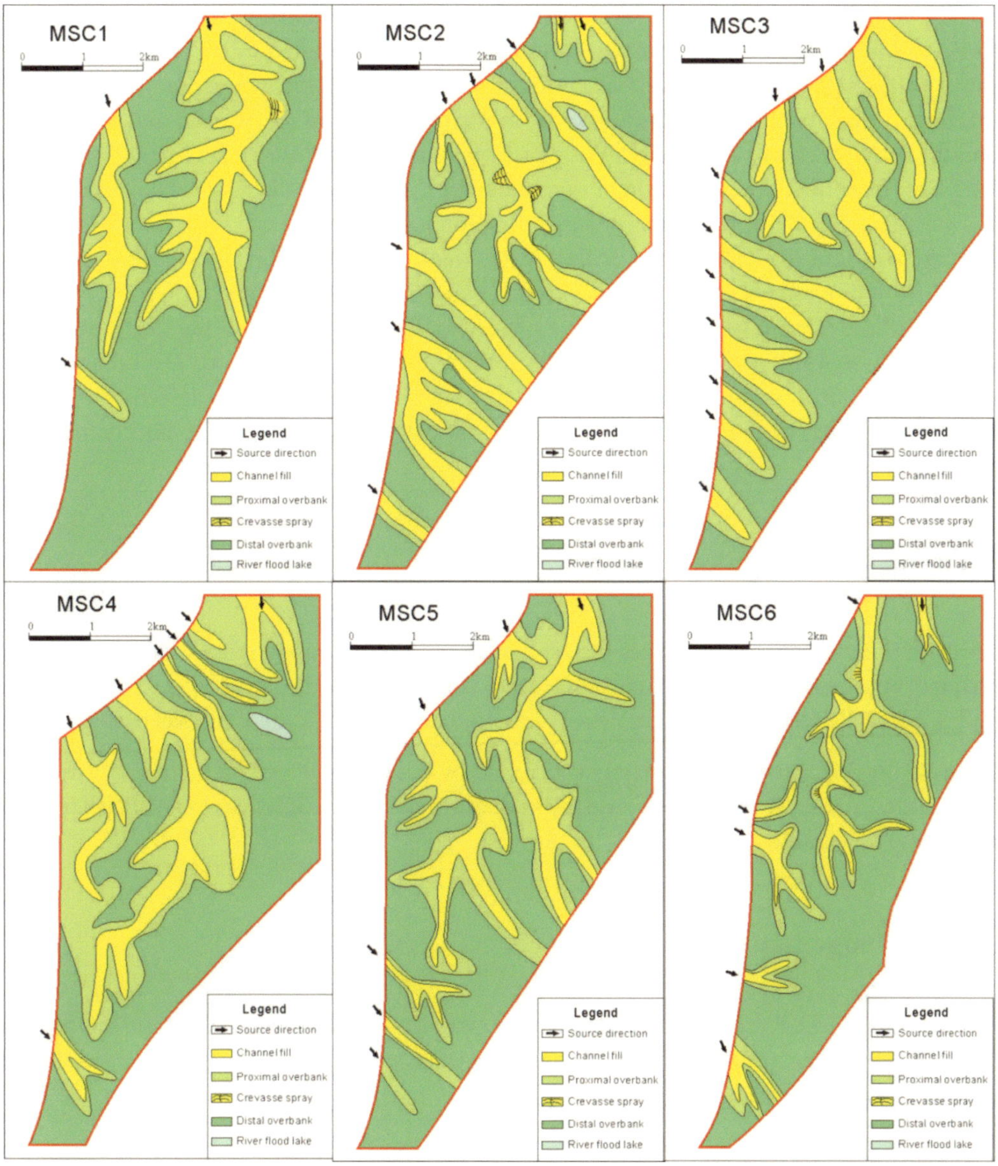

Figure 12. Sedimentary facies maps of the middle-term base level cycles of the target interval of W79 Block (after [19]).

overall coarsening-upward trend, revealing that the environment varied from a relatively higher accommodation condition to a relatively lower accommodation condition. This vertical change indicates that the entire LSC1 was formed during base level falling or an A/S decreasing period and is, in fact, a semi-cycle of the long-term scale. The vertical characteristic of LSC1 might reveal a relatively stable tectonic setting where subsidence tends to be slow and the depositional process is stable with abundant sediment supply.

LSC2, the second long-term base level cycle, composed of MSC5 to MSC6, approximately 150 to 190 m long, is the upper part of the target interval and is limited by SB1 at the top. In the lower part of this long-term base level cycle, similar to that of the upper part of LSC1, channel fill deposition is commonly found apart from the proximal and distal overbank deposition, and the connectivity of the sandbody is relatively high. In contrast, in the upper part of this long-term base level cycle, proximal overbank deposition, and distal overbank deposition are most commonly found, whereas channel fill deposition is sporadically found. The sedimentary stratigraphy shows an overall upward-fining trend, revealing that the environment varied from a relatively lower accommodation condition to a relatively higher accommodation condition. Similar to LSC1, the entire LSC2 was also a semi-cycle of the long-term scale, whereas it was formed during a base level rising or A/S increasing period. The vertical characteristics of LSC2 might reveal an active tectonic setting where subsidence tends to be fast and the base level rises rapidly, which makes the ratio of the channel to overbank change from high to low.

8. Discussions and conclusions

Through the regional correlation of the lithofacies within different depositional systems, this chapter proposes a high-resolution stratigraphic framework, including 2 long-term base level cycles, 6 middle-term base level cycles, and more than 58 short-term base level cycles.

The base level fluctuations are mainly controlled by the regional tectonic setting. Active subsidence stages tend to make base level rising semi-cycles, while relative stable stages tend to make base level falling semi-cycles.

Acknowledgements

Financial support was provided by the Fundamental Research Funds for Central Universities (2015KJJCB11). The authors also acknowledge permission by SINOPEC Zhongyuan Oilfield Company to publish this paper.

Author details

Jingzhe Li[1,2] and Jinliang Zhang[1*]

*Address all correspondence to: jinliang@bnu.edu.cn

1 College of Resources Science & Technology, Beijing Normal University, Beijing, China

2 School of Geosciences and Engineering, Sun Yat-sen University, Guangzhou, China

References

[1] Posamentier HW, Vail PR. Eustatic controls on clastic deposition II – sequence and systems tract models. In: Wilgus CK, Hastings BS, Kendall CGStC, Posamentier HW, Ross CA, Van Wagoner JC. (Eds.), Sea Level Changes – An Integrated Approach. Special Publication, Society of Economic Paleontologists and Mineralogists (SEPM), 1998;**42**;125-154

[2] Wright VP, Marriott SB. The sequence stratigraphy of fluvial depositional systems: The role of floodplain sediment storage. Sedimentary Geology. 1993;**86**(3-4):203-210

[3] Shanley KW, McCabe PJ. Alluvial architecture in a sequence stratigraphic framework: A case history from the Upper Cretaceous of southern Utah, USA. Quantitative Modeling of Clastic Hydrocarbon Reservoirs and Outcrop Analogues. Special Publication. 1993; **15**:21-55

[4] Shanley KW, McCabe PJ. Perspectives on the sequence stratigraphy of continental strata. AAPG Bulletin. 1994;**78**(4):544-568

[5] Schumm SA. River response to baselevel change: Implications for sequence stratigraphy. The Journal of Geology. 1993;**101**(2):279-294

[6] Olsen T, Steel R, Hogseth K et al. Sequential architecture in a fluvial succession; sequence stratigraphy in the Upper Cretaceous Mesaverde Group, Prince Canyon, Utah [J]. Journal of Sedimentary Research, 1995;**65**(2b):265-280

[7] Currie BS. Sequence stratigraphy of nonmarine Jurassic–Cretaceous rocks, central Cordilleran foreland-basin system. Geological Society of America Bulletin. 1997;**109**(9): 1206-1222

[8] Martinsen OJ, Ryseth A, Helland Hansen W, et al. Stratigraphic base level and fluvial architecture: Ericson Sandstone (Campanian), Rock Springs Uplift, SW Wyoming, USA. Sedimentology. 1999;**46**(2):235-263

[9] Blum MD, Törnqvist TE. Fluvial responses to climate and sea-level change: A review and look forward. Sedimentology. 2000;**47**(s1):2-48

[10] Catuneanu O. Sequence stratigraphy of clastic systems: Concepts, merits, and pitfalls. Journal of African Earth Sciences. 2002;**35**(1):1-43

[11] Holbrook J, Scott RW, Oboh-Ikuenobe FE. Base-level buffers and buttresses: A model for upstream versus downstream control on fluvial geometry and architecture within sequences. Journal of Sedimentary Research. 2006;**76**(1):162-174

[12] Zecchin M, Baradello L, Brancolini G, et al. Sequence stratigraphy based on high-resolution seismic profiles in the late Pleistocene and Holocene deposits of the Venice area. Marine Geology. 2008;**253**(3):185-198

[13] Fanti F, Catuneanu O. Fluvial sequence stratigraphy: The Wapiti Formation, west-central Alberta, Canada. Journal of Sedimentary Research. 2010;**80**(4):320-338

[14] Holbrook JM, Bhattacharya JP. Reappraisal of the sequence boundary in time and space: Case and considerations for an SU (subaerial unconformity) that is not a sediment bypass surface, a time barrier, or an unconformity. Earth-Science Reviews. 2012;**113**(3):271-302

[15] Zhang J, Li J, Liu S, et al. Sedimentology and sequence stratigraphy of the second member of Shuangyang Formation, Y45 Block, Moliqing oilfield, Yitong Basin, China. Arabian Journal of Geosciences. 2015;**8**(9):6697-6707

[16] Catuneanu O. Principles of Sequence Stratigraphy. Elsevier, Amsterdam. 2006. pp.222-229

[17] Miall AD. Stratigraphy: The Modern Synthesis. Springer, Heidelberg; 2016. p. 311-370

[18] Smith DG, Smith ND. Sedimentation in anastomosed river systems; examples from alluvial valleys near Banff, Alberta [J]. Journal of Sedimentary Research, 1980;**50**(1):157-164

[19] Li J, Zhang J, Liu S, et al. Sedimentology and sequence stratigraphy of the Paleogene lower second member of the Shahejie Formation, W79 Block, Wenliu Oilfield, Bohai Bay Basin, China. Russian Geology and Geophysics. 2016;**57**(6):944-957

[20] Weissmann GS, Hartley AJ, Nichols GJ, et al. Fluvial form in modern continental sedimentary basins: Distributive fluvial systems. Geology. 2010;**38**(1):39-42

[21] Weissmann GS, Mount JF, Fogg GE. Glacially driven cycles in accumulation space and sequence stratigraphy of a stream-dominated alluvial fan, San Joaquin Valley, California, USA. Journal of Sedimentary Research. 2002;**72**(2):240-251

[22] Gibling MR. Width and thickness of fluvial channel bodies and valley fills in the geological record: A literature compilation and classification. Journal of Sedimentary Research. 2006;**76**(5):731-770

[23] Gawthorpe RL, Leeder MR. Tectono-sedimentary evolution of active extensional basins. Basin Research. 2000;**12**(3-4):195-218

[24] Posamentier HW. Lowstand alluvial bypass systems: Incised vs. unincised. AAPG Bulletin. 2001;**85**(10):1771-1793

[25] Koss JE, Ethridge FG, Schumm SA. An experimental study of the effects of baselevel change on fluvial, coastal plain and shelf systems [J]. European Journal of Medicinal Chemistry, 1994;**21**(1):65-69

[26] Holbrook J, Schumm SA. Geomorphic and sedimentary response of rivers to tectonic deformation: A brief review and critique of a tool for recognizing subtle epeirogenic deformation in modern and ancient settings. Tectonophysics. 1999;**305**(1):287-306

[27] Blum MD. Genesis and architecture of incised valley fill sequences: a Late Quaternary example from the Colorado River, Gulf Coastal Plain of Texas. In: Weimer, P., Posamentier, H.W. (Eds.), Siliciclastic sequence stratigraphy: recent developments and applications. Memoir, American Association of Petroleum Geologists, 1994;**58**:259-283

[28] Sylvia DA, Galloway WE. Morphology and stratigraphy of the late Quaternary lower Brazos valley: Implications for paleo-climate, discharge and sediment delivery [J]. Sedimentary Geology, 2006;**190**(1-4):159-175

[29] Jervey M T. Quantitative geological modelling of siliciclastic rock sequences and their seismic expression [J]. 1988;**42**:47-69

[30] Posamentier HW, Jervey MT, Vail PR. Eustatic controls on clastic deposition I — conceptual framework. In: Wilgus CK, Hastings BS, Kendall CGStC, Posamentier HW, Ross CA, Van Wagoner JC (Eds.), Sea Level Changes — An Integrated Approach. Special Publication, Society of Economic Paleontologists and Mineralogists (SEPM), 1988;**42**:110-124

[31] Leckie DA, Boyd R. Towards a nonmarine sequence stratigraphic model. American Association of Petroleum Geologists Annual Convention, Salt Lake City, 11-14 May 2003. Official Program, 2003;**12**: p. A101

[32] Li J, Liu S, Zhang J, et al. Architecture and facies model in a non-marine to shallow-marine setting with continuous base-level rise: An example from the Cretaceous Denglouku Formation in the Changling Depression, Songliao Basin, China. Marine and Petroleum Geology. 2015;**68**:381-393

[33] Holbrook JM. Complex fluvial response to low gradients at maximum regression: A genetic link between smooth sequence-boundary morphology and architecture of overlying sheet sandstone. Journal of Sedimentary Research. 1996;**66**(4):713-722

[34] Burns BA, Heller PL, Marzo M, et al. Fluvial response in a sequence stratigraphic framework: Example from the Montserrat fan delta, Spain. Journal of Sedimentary Research. 1997;**67**(2):311-321

[35] Cant DJ. Subsurface Facies Analysis. Facies Models, Response to Sea Level Changes. Canada: Geol. Assoc.; 1992. p. 27-45

[36] Posamentier HW, Allen GP. Siliciclastic sequence stratigraphy: Concepts and applications. In: SEPM (Society for Sedimentary Geology) Tulsa. 1999

[37] Parasnis DS. Seismic methods. Springer, Heidelberg; 1986. p. 250-317

[38] Hardage BA. Vertical seismic profiling. The Leading Edge. 1985;**4**(11):59

[39] Posamentier HW, Kolla V. Seismic geomorphology and stratigraphy of depositional elements in deep-water settings. Journal of Sedimentary Research. 2003;**73**(3):367-388

[40] Zuo Y, Qiu N, Zhang Y, et al. Geothermal regime and hydrocarbon kitchen evolution of the offshore Bohai Bay Basin, North China. AAPG Bulletin. 2011;**95**(5):749-769

[41] Li H, Peng S. Study on Sedimentary Microfacies During the Middle-Late Development Stages of Wen 72 Block in Wenliu Oilfield. 2007. p. 4-710

[42] Jiang S. Potential Evaluation and Classification of Favorable Areas in the Bohai Bay Basin. China University of Petroleum (EastChina), 2014.

[43] Dai Y. Study of the High-Resolution Sequence Stratigraphy and Sedimentary Microfacies in Ming 1 Fault Block of Wenming Village Oilfield. Ocean University of China; 2012

[44] Wang S. Fine Reservoir Description of Block Wen25 in Wenliu Oilfield. Ocean University of China; 2011

[45] Fu J. Paleogeographic Background and Sedimentary Facies Study on the Lower of the Third Member of the Shahejie Formation of Paleogene in Dongpu Sag. Beijing: China University of Geosciences; 2008

[46] Fu M, Li W, Jia R, et al. Study of progressive exploration and development in Wenliu oilfield. Special Oil & Gas Reservoirs. 2005;**3**:3

[47] Hu M, et al. Sedimentary model and microfacies division of Wenliu Area. Inner Mongolia Science Technology and Economics, 2014;**8**:8-64

[48] Miall AD. Architectural-element analysis: A new method of facies analysis applied to fluvial deposits. Earth-Science Reviews. 1985;**22**(4):261-308

[49] Holz M, Vieira PE, Kalkreuth W. The early Permian coal-bearing succession of the Paraná Basin in southernmost Brazil: Depositional model and sequence stratigraphy. Revista Brasileira de Geociencias. 2000;**30**(3):420-422

[50] Zhang J, Xie J. Reservoir Sedimentary Facies Model. Beijing: Petroleum Industry Press; 2008

[51] Vakulenko LG, Aksenova TP, Yeltsov IS, et al. A lithofacies description of Jurassic sediments in the south of the Predyenisei petroleum subprovince, West Siberia. Russian Geology and Geophysics. 2010;**51**(4):329-338

[52] Srivastava AK, Mankar RS. Lithofacies architecture and depositional environment of Late Cretaceous Lameta Formation, central India. Arabian Journal of Geosciences. 2015;**8**(1):207-226

Chapter 4

Seismic Stratigraphic Features of the Late Miocene-Present Unconformities and Related Seismic Units, Northern Offshore Taiwan

Jih-Hsin Chang, Eason Yi-Cheng Yang, Ho-Han Hsu, Chih-Chieh Su, Char-Shine Liu, Shye-Donq Chiu, Yu-Fang Ma, Yuan-Wei Li, Yen-Chun Lin and Jen-Sen Shen

Additional information is available at the end of the chapter

http://dx.doi.org/10.5772/intechopen.70819

Abstract

We investigate the seismic stratigraphic features offshore northern Taiwan by using newly collected multichannel seismic data. Two significant regional unconformities U1 and U2 have been identified, which further subdivide the sedimentary sequence into three seismic units as SU I, SU II, and SU III. The lowermost seismic unit SU I is a pre-late Miocene sequence, while the middle and upper seismic unit SU II and SU III result from the interactions between the rapid fault-controlled subsidence and the stable thermal-controlled subsidence. We consider that the present-day offshore northern Taiwan is under a post-collisional state and the unconformities U1 and U2 represent a response to the mountain collapse and to the cessation of the regional volcano-tectonic activities. It is not until 1.5 Ma that northern offshore Taiwan became a post-collisional basin and started to receive sediments, with a rapid fault-controlled subsidence. Afterward, the basin became dominated by a stable thermal-controlled subsidence at 0.2 Ma. Although the main volcano-tectonic activities in the northern offshore Taiwan are ceased, modern geophysical and geochemical investigations have suggested that the tectonism and the volcanism are still active and represent potential threatening geohazard.

Keywords: seismic reflection, orogeny, mountain collapse, northern offshore Taiwan

1. Introduction

In addition to gravimetric and magnetic data, the reflection seismic data are powerful tools to understand the subsurface geological features and to determine the nature of the sedimentary basins. In terms of regional stratigraphy, they provide not only the thickness and the distribution of the sedimentary sequences but also the contact relationships between the sedimentary sequences, representing stratigraphic lap-out and geological unconformities. In addition to sea level change, tectonic events are generally accepted as a general cause for these seismic unconformities, indicating the basin formation mechanisms directly. In this study, we aim at analyzing the dominant tectonic events and the present-day tectonic setting of northern offshore Taiwan based on newly collected marine reflection seismic data, which reveal the subsurface stratigraphic features.

The northern offshore Taiwan is located at the junction among the southernmost part of the East China Sea, the south-western extension of Okinawa Trough and the northern tip of the Taiwan orogenic belt (**Figure 1**). The northern offshore area of Taiwan is surrounded by different geological units highlighting that several basins have influenced its tectono-sedimentary evolution. It could be the part of the post-rift stage of the Paleogene rift basin in the East China Sea [1, 2]. It may also be proposed that it is dominated by relict back-arc basins, which were controlled by a progressive eastward migrating subduction of the Pacific plate and ended up in the present Okinawa Trough [3–5]. The progressive southward migration of the Taiwan orogenic belt could be a practical mechanism for basin formation as well [6–9]. Recently, strike-slip motion along the East Asia continental margin is considered to play a role on the basin evolution of the East China Sea [10]. A re-appraisal of which basin formation mechanism is more dominant for the Neogene basin development in the northern offshore Taiwan is thus required.

This chapter provides new seismic stratigraphic information on the Neogene tectonic setting in the northern offshore Taiwan through seismic interpretation. The stratigraphic architecture of the Neogene sedimentary sequences in the northern offshore Taiwan has been reconstructed through the geological interpretation of high-resolution seismic profiles. Here, we examine two high-resolution reflection seismic profiles of different orientations northern offshore Taiwan (**Figure 2**). One of the profiles runs in NE-SW direction, showing a northward dipping sequence. The other one, on the other hand, runs in NW-SE direction, showing a series of tilted fault blocks. Most significantly, both profiles feature obvious, angular unconformities. The geological and tectonic significance of these unconformities is discussed in this study.

2. Geological backgrounds

Connecting Japan to the north and Taiwan to the south, the East China Sea is a marginal sea over a broad continental shelf (**Figure 1a**). The basin of the East China Sea shelf, also known as East China Sea Shelf Basin, is the largest Cenozoic sedimentary basin of the East Asia continental margin [2]. There are several sub-basins at the southern end of the East China Sea Shelf Basin (**Figure 1b**) [11]. These sub-basins were formed in Paleogene, filled with the syn-rift sedimentary sequence and covered by Neogene post-rift sedimentary sequence [1, 12].

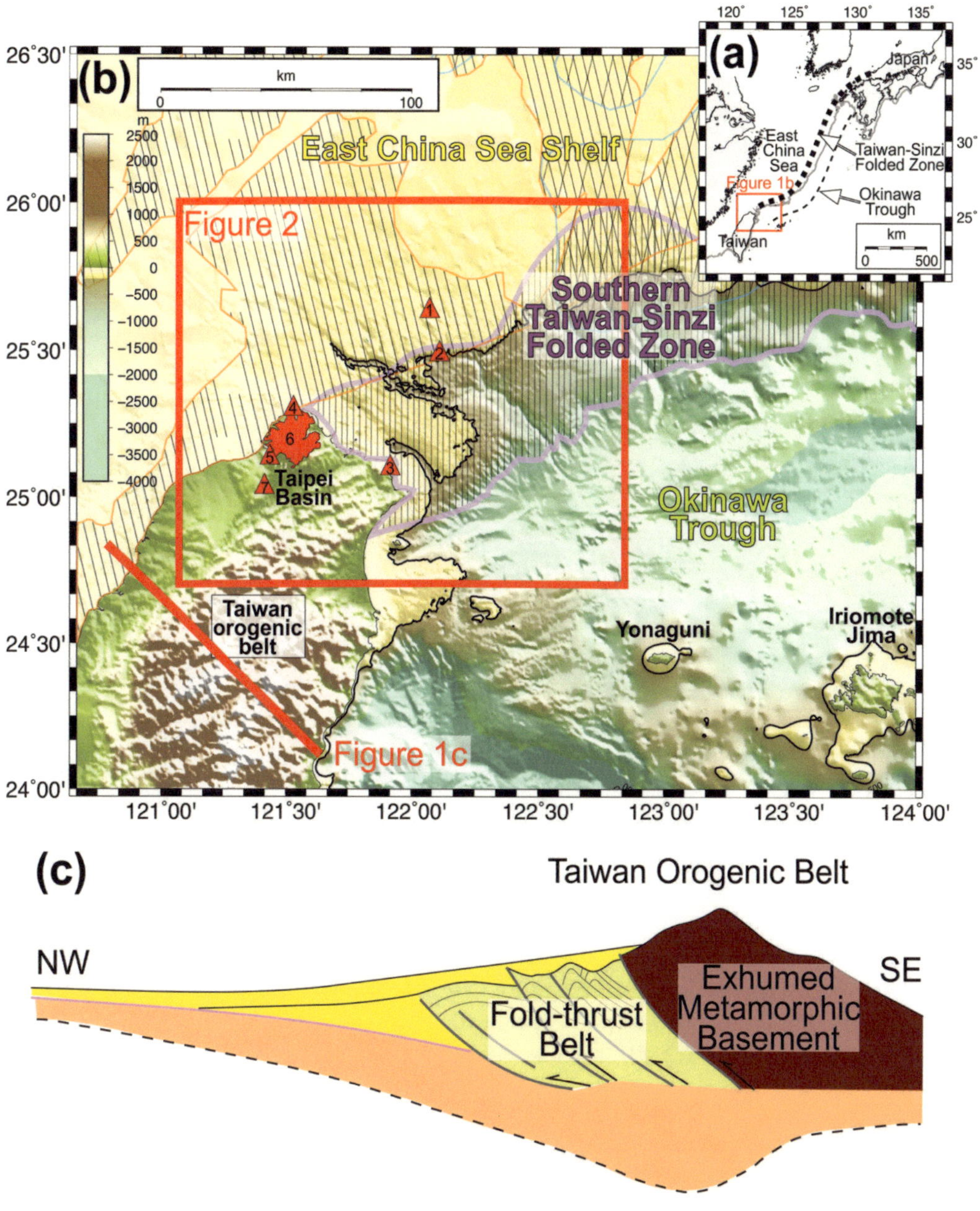

Figure 1. Geological settings and basin location in the study area. (a) Regional setting map showing the approximate location of the East China Sea, of the Okinawa Trough (thin dashed line) and of the Taiwan-Sinzi Folded Zone (thick dashed line). (b) Bathymetric map showing the location of the Paleogene rift basins identified by seismic data (backslash area) and back-arc basin identified by magnetic data (slash area), southern Taiwan-Sinzi Folded Zone (vertical bar line), and Northern Taiwan Volcanic Zone (red area and triangles. (1) Pengjia Islet, (2) Mianhua Islet, (3) Keelung Volcanic Group, (4) and (5) Kuanyinshan, (6) Tatun Volcanic Group, and (7) Tsaolingshan). The black thin lines indicate 200 m contour, and the Red thick line indicates the location of the schematic profile shown in **Figure 1c**. (c) Sketch profile across the Taiwan orogenic belt at its culminating stage The green and yellow area indicate the deformed Tertiary-present sequence, respectively.

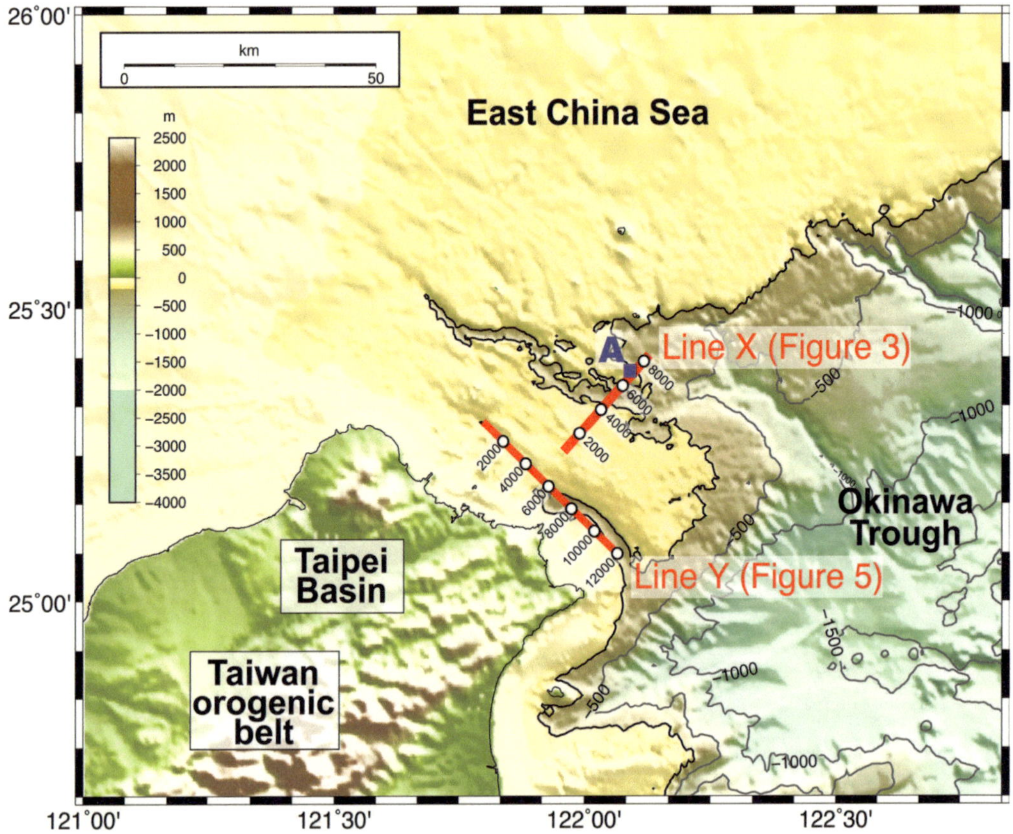

Figure 2. Bathymetric map with location of seismic lines. Red lines indicate the locations of the selected seismic profiles. White circles along the red lines indicate the locations of the shot points. Blue square A indicates the location of the exploration wells.

The Okinawa Trough is a long, N-S trending that connects Japan and Taiwan (**Figure 1**). It is a back-arc basin formed by extension within the continental lithosphere landwards of the Ryukyu trench-arc system [13, 14]. The initial opening of the northern and middle part of the current Okinawa Trough occurred during the Miocene, while it delayed until the Pleistocene in the southern part of the Trough [15, 16]. It is also believed that this back-arc basin may be suited even westwards before Miocene, controlling the emplacement and the tectonic setting of the East China Sea Shelf Basins [3, 5].

The Taiwan is located between the large marginal seas of the East China Sea and the South China Sea and includes a young orogenic belt, formed by Late Miocene collisional events in the SE Eurasian continent (**Figure 1b** and **c**) [17, 18]. In addition to metamorphic basement rocks, the body of the Taiwan orogenic belt is mainly composed of metamorphic basement rocks and deformed Tertiary rift basins, including Paleogene syn-rift and Neogene post-rift sedimentary sequences [19]. Since the Taiwan orogenic belt is moving southward, different evolutionary stages of the orogenic belt features along the trending of Taiwan have been distinguished [20, 21]. The central part of the Taiwan, which is characterized by high mountain peaks of Taiwan mountain ranges, is bearing the culmination of the collisional activity (**Figure 1b**) [22, 23]. Accordingly, the northward descending topography of the mountain range represents the northward decline of the collision [6].

The Taiwan-Sinzi Fold Zone is another significant geological unit in this area (thick dashed line in **Figure 1a**; purple area in **Figure 1b**) [8, 9, 24–28]. It runs approximately along the shelf break, trending parallel to the Okinawa Trough and representing a structural high separating the East China Sea Basin to the west and the Okinawa Trough to the east. In the northern and middle part, the Taiwan-Sinzi Folded zone is characterized by structural highs of the acoustic basement, while it shows folded structures and tilted blocks in the southern part, interpreted as the earliest Taiwan orogenic belt [7, 8, 29].

Northern Taiwan area is also characterized by the Quaternary-present magmatic activities, which is also known as Northern Taiwan Volcanic Zone (NTVZ). The NTVZ is composed of several groups of the volcanic fields, including the earliest (before 2.6 Ma) Tatun Volcanic Group, Mianhua Islet, and Sekibisho; later (2~1 Ma) Tatun Volcanic Group, Pengjia Islet and Keelung Volcanic Group; and the latest (after 1Ma) Tsaolingshan, Kuanyinshan, and Kobisho (**Figure 1b**). They occur probably in association with the westernmost part of the Ryukyu Arc or in response to the extensional magmatism of the northern Taiwan mountain range [6, 30, 31].

3. Data acquisition and processing

In this study, we used the multichannel reflection seismic data collected by the Taiwanese research vessel *Ocean Researcher II*. The source energy was shot via a 210-cubic inch GI gun with shot interval of 37.5 m. The reflection seismic data were acquired through a 24-channel, 150-meter long streamer, with a 1-ms sampling rate. The reflection seismic data were processed by the ProMAX and KINGDOM software at the Institute of Oceanography, National Taiwan University. The data-processing procedures include trace editing, geometry set-up, band-pass filtering, amplitude compensation, predictive deconvolution, velocity analysis, normal move-out correction, stacking, water velocity F-K time migration, and water column mute. Bathymetric dataset is a 500-m gridded data, produced by compiling shipboard and global dataset [32]. The bathymetric charts are prepared by the GMT [33].

4. Results

4.1. Line X

We interpret the seismic data and depict stratigraphic boundaries, mainly on the basis of the concepts carried out by Vail and Mitchum [34] and Mitchum et al. [35]. Line X runs NE-SW, extending approximately 30 km and perpendicularly to the coastline of Taiwan (**Figure 2**).

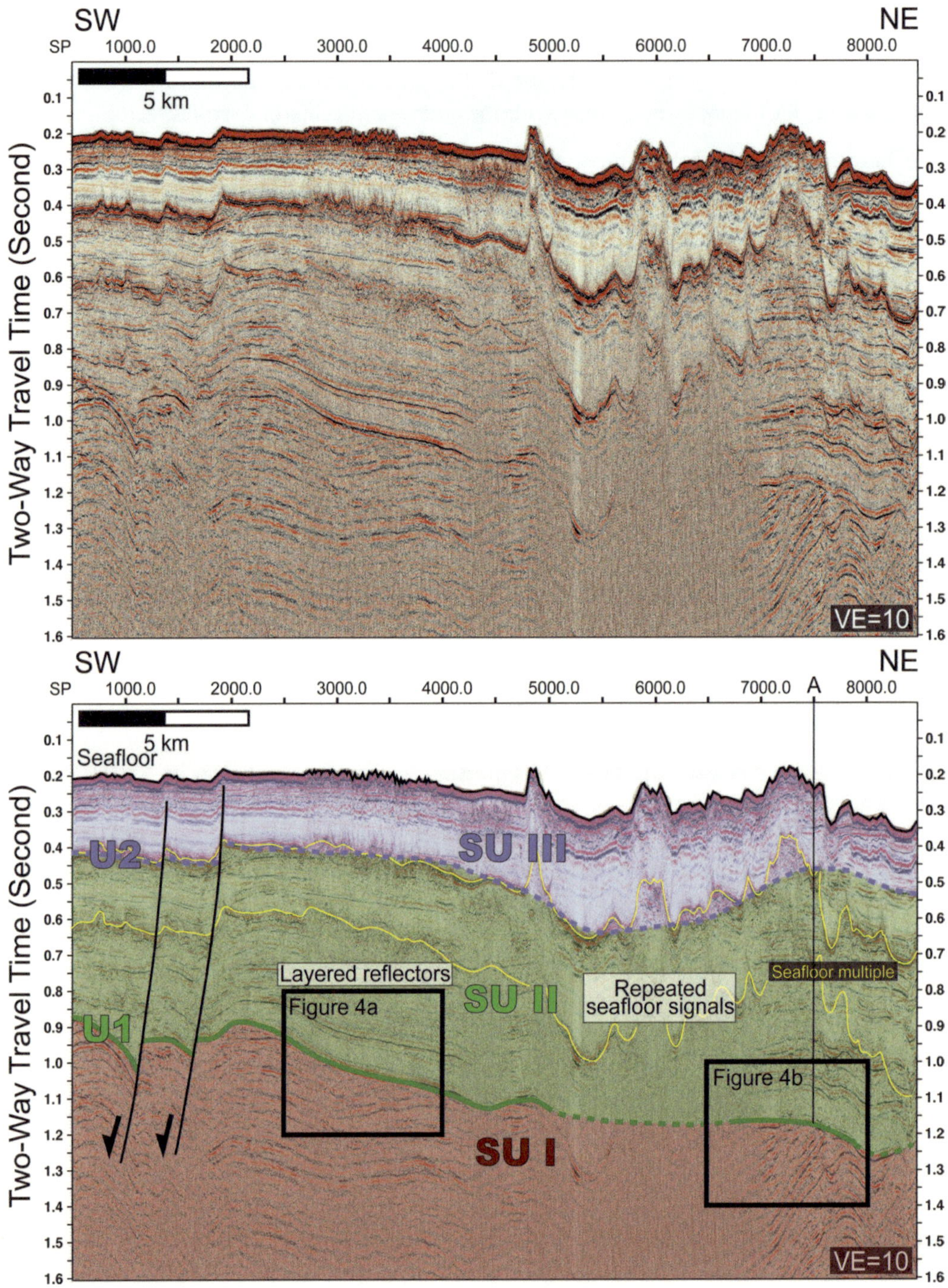

Figure 3. Selected reflection seismic profile along Line X. See **Figure 2** for the location.

The most of line X is located on the continental shelf, and the central and northeastern part of the Line X have the bathymetry deeper than 200 m (**Figures 2** and **3**). A drilling site at the NE part of the survey line X, providing geological information and chronostratigraphic controls in study area [8, 9].

A seismic boundary namely U1 at about 1 second Two-Way Travel Times (twt) has been observed (**Figure 3**, marked as a green line), extending in the whole profile. In the middle-south part of the profile X (SP 2500-4000; **Figure 4a**), the reflectors beneath the U1 show very apparent termination against the U1, of which we interpreted as erosional truncation. In the northeastern part of the profile (SP 6500-8000; **Figure 4b**), the termination relationship between the U1 and the underlying reflectors shows erosional truncation as well. Collectively, we suggest that U1 is an unconformable surface. At the southwestern part of the profile, the unconformity U1 is located around 0.85 s (twt) (**Figure 3**), while the unconformity U1 becomes deeper to around 1.2 s (twt) at the north-eastern end of the profile, showing a gently descending trend toward north-east.

The characteristics of the seismic reflectors above the unconformity U1 are not always consistent along the Line X. In the central-south and northernmost part of the profile (**Figure 3**,

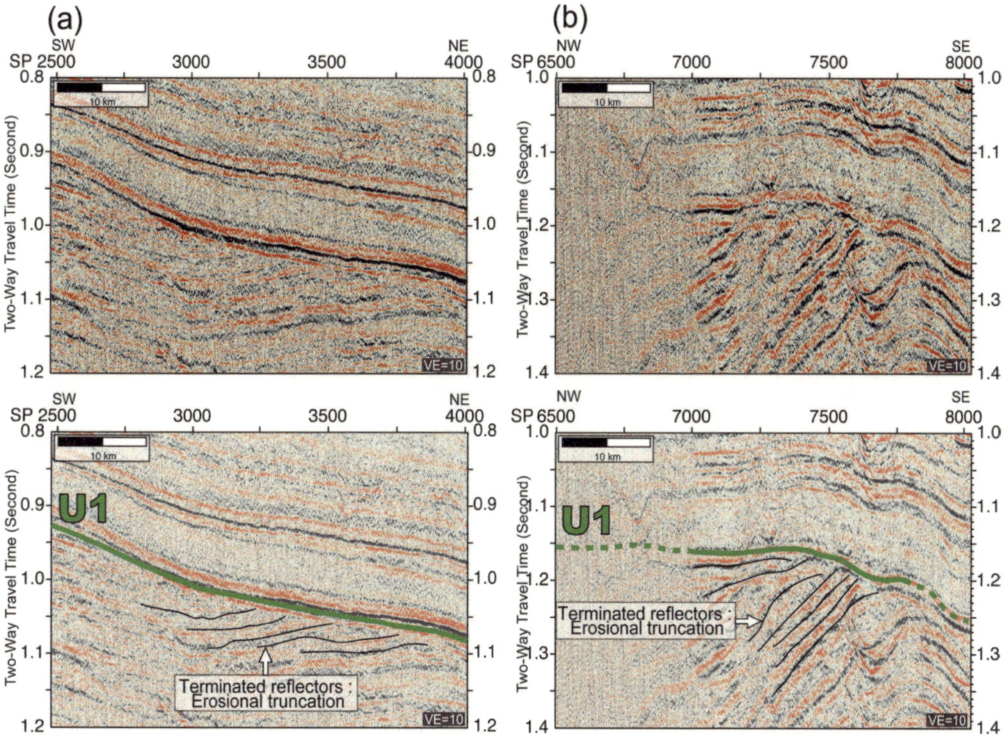

Figure 4. Selected seismic sections showing the termination feature of the U1. See **Figure 3** for locations.

SP 500-5000; 7000-8500), it is dominated by layered reflectors. However, in the middle-north part of the profile (**Figure 3**, SP 5000-7000), it shows strong seafloor reverberation above 1 s (twt) and an area of reflection-free below 1 s (twt). It seems that the strong seafloor reverberation has intervened between layered reflectors to both southern and northern ends of the seismic profile. We interpret that this reflection feature is plausible to be the seismic response of very hard and solid seafloor material, probably crystallization or extrusive rocks.

4.2. Line Y

Line Y is located near offshore northern Taiwan, and runs NW-SE, parallel to the coastline of the Taiwan (**Figure 2**). It extends around 40 km along dip of the continental shelf, showing even no drastic change in bathymetry (**Figures 2** and **4**). There are some gentle bathymetric reliefs associated with dominant fault structures (indicated by write triangles in **Figure 5**; SP 4300-4500, 5900-6100, 8300-8600).

In the north-western part of the seismic profile Y (SP 400-1700; **Figure 6a**), the truncated reflectors are tilted, suggesting the occurrence of tectonic and/or erosional events. In addition, the

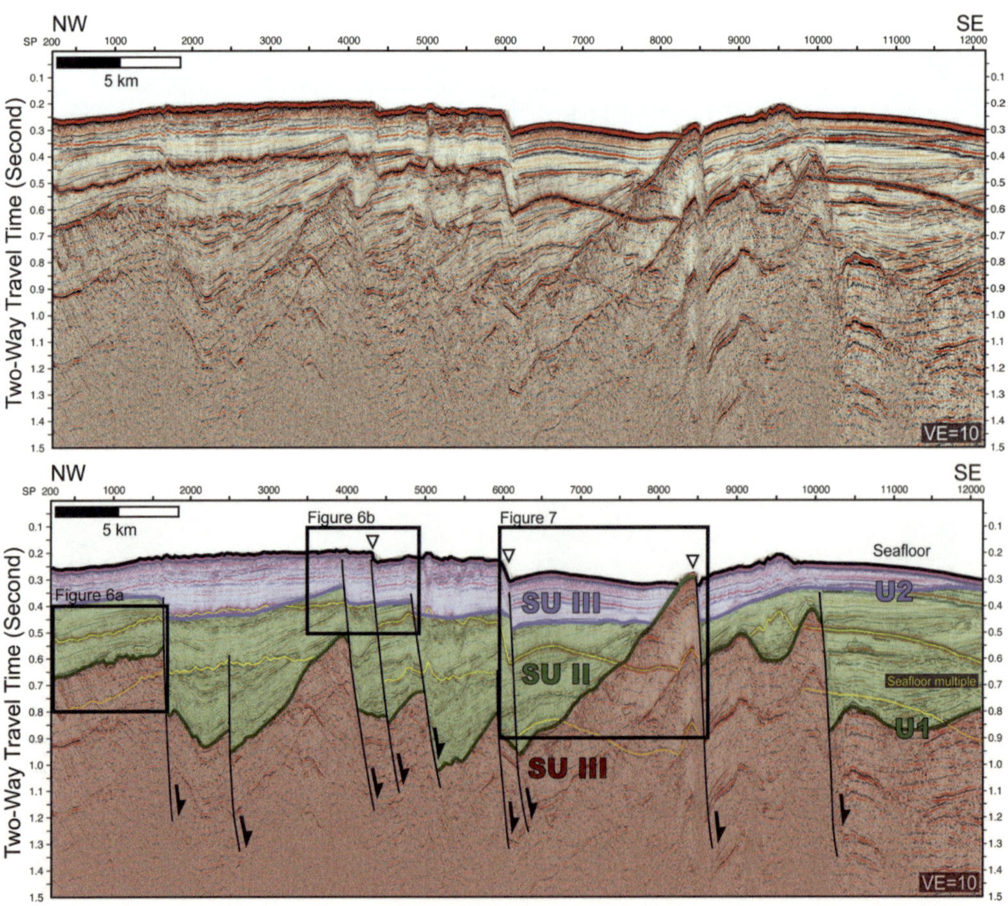

Figure 5. Selected reflection seismic profile along Line Y. See **Figure 2** for the location.

http://dx.doi.org/10.5772/intechopen.70819

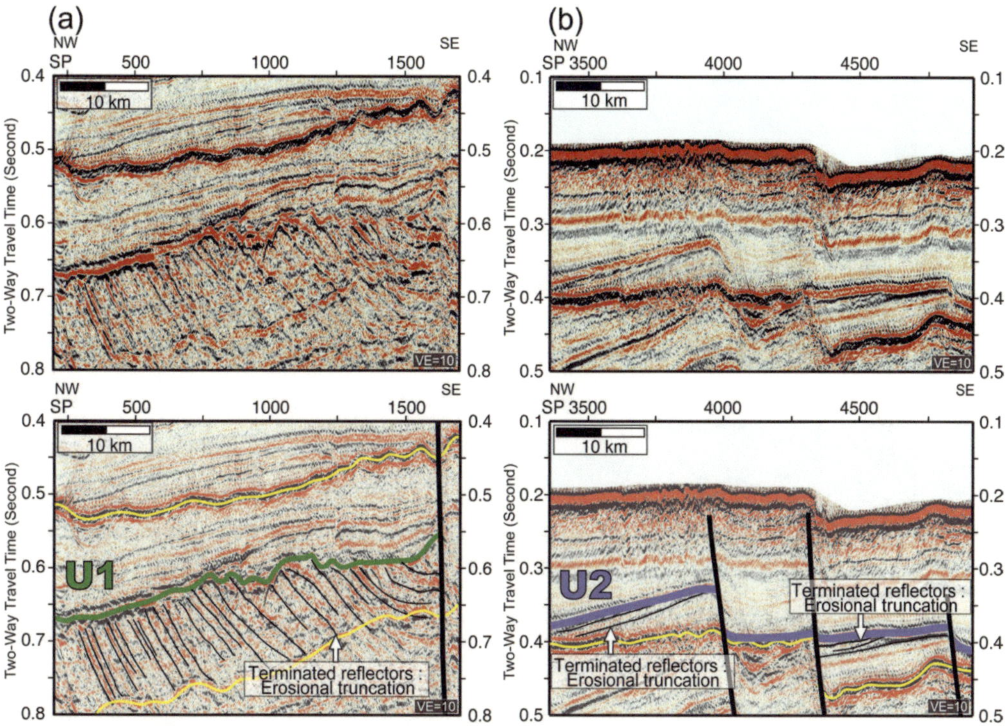

Figure 6. Selected seismic sections showing the termination feature of the U1 and U2. See **Figure 5** for locations.

U1 in the seismic profile Y also serves as a top surface of the dominating tilted fault blocks. In this way, the U1 is of regional importance, and the distribution of the U1 reflects the local tilting of the fault blocks.

We also identified another seismic boundary U2 in the seismic profile Y (**Figure 5**, marked as purple lines), although it often coexists with seafloor multiple signals (**Figure 5**, SP 1700-3000 and 5000-6000). In the middle-south part of the profile (SP 3500-5000; **Figure 6b**), the U2 is identified at around 0.3-0.4 s (twt), serving as a termination surface of underlying updipping reflectors. On the basis of on the parallelism of terminated reflectors [35], we interpret an erosional truncation relationship between the U2 and terminated reflectors. In the middle-north part of the profile (SP 6000-8700; **Figure 7**), the parallelism of terminated reflectors and the feature of erosional truncation become much obvious. We interpret the U2 as an unconformity that may mark a tectonic event as the unconformity U1.

We note that both the unconformities U1 and U2 do crop out at seafloor in the seismic profile Y (**Figure 6**). The unconformity U1 crops out at the crest of a local structural high, along with a fault-block bounding fault (**Figure 7**, SP 8200-8500). The U2 crops out at southeasternmost of the profile (**Figure 7**, SP 12000). These may be the result of the interaction between the degree of fault block rotation and local sediment discharge. In addition to the U1 and U2 in the seismic profile Y, we observe a strong and continuous reflector shown between the clear features of the U1 and U2 (dashed line in **Figure 7**). Such reflector may also indicate the existence of the minor and local events, probably the halts of the fault block rotations.

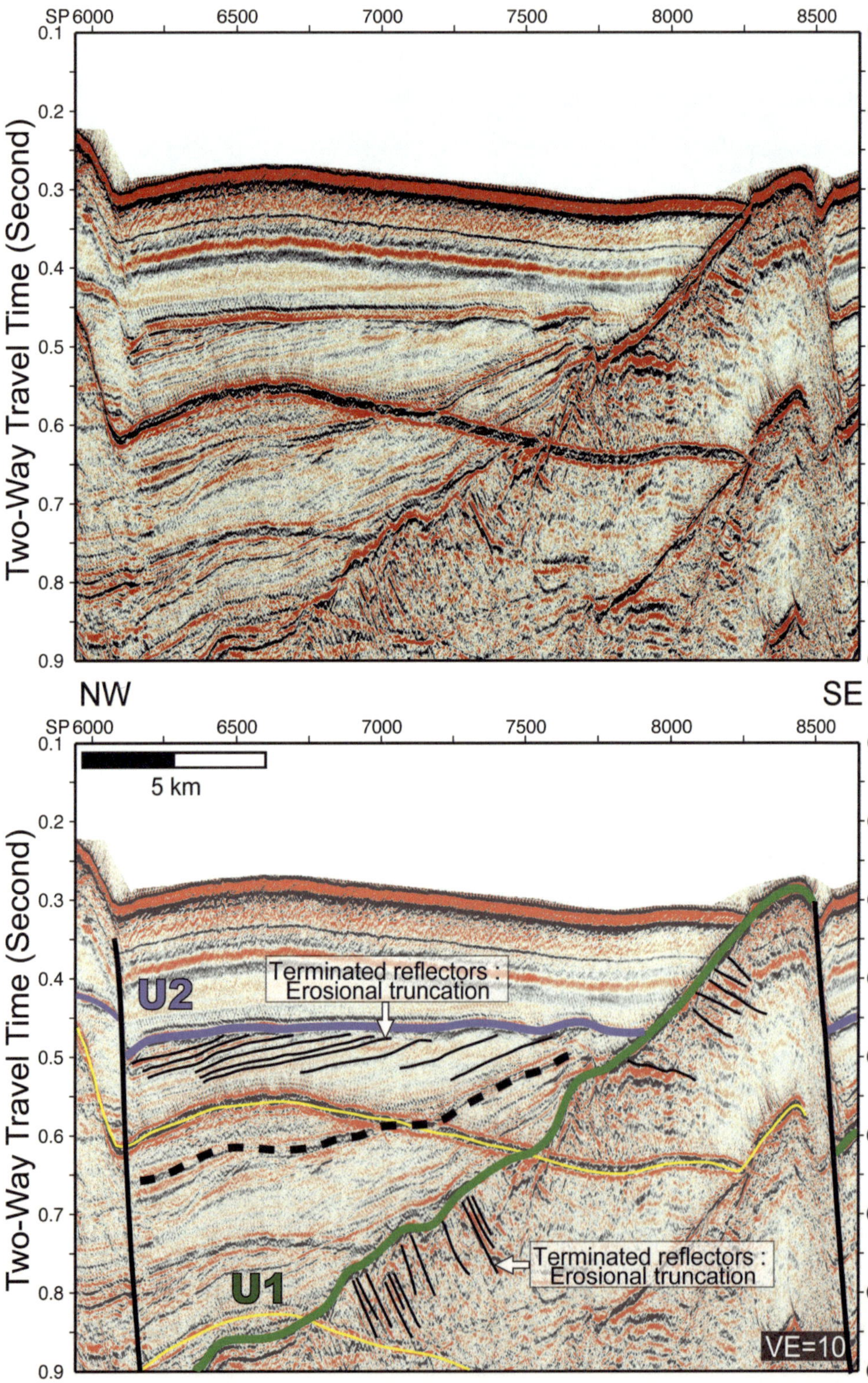

Figure 7. Selected seismic sections showing the termination feature of the U1 and U2. See **Figure 5** for locations.

4.3. Seismic units

On the basis of the U1 and U2, three seismic units are determined accordingly: the SU I, SU II, and SU III, which indicate the sedimentary sequence beneath the U1, the sedimentary sequence between the U1 and U2, and the sedimentary sequence above U2, respectively (**Figures 3** and **5**). In the seismic profile Y, we note that the thickness of the SU II varies along the profile, mainly depending on the relief of the underlying fault structures (**Figure 5**), while the thickness of the SU III change slightly and diminish toward the southeast end of the seismic profile Y (**Figure 5**, SP 10000-12000). The thickness variation of the SU II and SU III may indicate the change in subsidence patterns and rates, probably reflecting the change in controlling factor of subsidence [36, 37, 38].

5. Discussion and conclusion

5.1. Age of U1 and U2

In our seismic data, the seismic boundaries U1 and U2 are remarkable unconformities, which show similarity in the truncation relationship with underlying strata. The ages of the unconformities U1 and U2 formation are thus significant, revealing more information on regional tectonic events. Based on drilling results [8], the strata overlying the unconformity U1 were dated back to approximately Pleistocene and the strata beneath the same unconformity were dated back to Late Miocene [8]. Accordingly, the unconformity U1 bears a hiatus of Late Miocene-Quaternary age, while there is still no available drilling information reporting the existence of the unconformity U2 up-to-date. In this way, the age of the unconformities U2 may be alternatively proposed by additional local Late Miocene-present tectonic events.

The Quaternary magmatism is also a significant feature in post-collisional environment in northern Taiwan, resulting in the NTVZ [30]. The ages of the NTVZ have been dated, mainly distributing from 2.8 to 0.2 Ma. During this period, three age groups of before 2.6 Ma, 2~1 Ma, and after 1 Ma have been identified [30]. We note that the first age group may represent the earliest signals of the post-collisional state [30]. Among the three groups, the second age group is mainly composed of the volcanic edifice formed at the Pengjia Islet (2.1 Ma), the Tatun Volcanic Group (1.5 Ma), and the Keelung Volcanic Group (1.4 Ma). We consider that, the formation of the unconformity U2 and deposition of the SU II are likely simultaneous to the second age group of the NTVZ [8]. Another point we note is that most of the NTVZ events may cease no later than 0.2 Ma [30], suggesting that 0.2 Ma may be a critical timing of tectonic environment change in the NTVZ area. Collectively, we suggest the unconformities U1 and U2 may represent the age of 1.5 Ma and 0.2 Ma, respectively.

5.2. Late Neogene basin of north offshore Taiwan

The formations of the unconformities U1 and U2 shall be related to the tectonic development of the Late Miocene-present sedimentary basin of northern offshore Taiwan, which has been considered as post-rift basin [1, 2], back-arc basin [3–5], post-collisional basin [6–8], and strike-slip basin [10]. Among the four competing models, the strike-slip motion has been proposed for long time, as it may be caused by the formation of the East China Sea Shelf Basin or the opening southern Okinawa Trough [9, 10]. However, the shape of the Late Neogene basin does not present as a rhombic shape reflecting strike-slip pull apart basins. In addition, the strike-slip structures have not been fully supported by mapping result from marine reflection seismic survey. The post-rift basin model is proposed as northern offshore Taiwan being considered as an extension of the East China Sea Shelf Basin. While in our study area, the post-rift shall have been ceased since the activities of both Taiwan orogenic belt to the south and the back-arc extension of the Okinawa Trough to the southeast are more immediate and are more likely to dominate. On the basis of the geographical connection of our study area with the Okinawa Trough and the Taiwan mountain range, the rest back-arc basin and post-collisional basin models are better candidates for the Late Neogene basin formation.

Our reflection seismic data show that U1 is an unconformity that truncated the underlying fault blocks and the tilting strata. We consider that those tilting strata may indicate a drastic uplift event prior to the formation of the unconformity U1. As a result, the back-arc basin model may be less likely to ascribe for strong tectonic uplift, once the back-arc basin is more dominated by crustal extension. In this way, we tend to agree with the post-collisional basin model. We suggest that the unconformity U1 represents an age mark of the post-collisional stage that started to receive sediments; the SU I were the strata deposited before the mountain collapse, and the SU II were deposited probably after the post-collisional basin had formed. The thickness variation of the SU II depends on the distribution of faults, indicating that SU II may be formed in association with fault-controlled subsidence [36, 37, 38]. The unconformity U2 shall be a regional tectonic event later in the Quaternary. Based on its erosional and relatively flattening feature upon tilting fault blocks, we propose that the unconformity U2 may be related to the cessation of the fault blocks rotation and change of regional subsidence rate. The thickness of the SU III does not vary greatly, indicating that the fault-controlled subsidence was followed by a relatively stable, probably a thermal-controlled subsidence [36, 37, 38].

5.3. Tectonic evolution and implication of post-collisional basin

Figure 8 is a sketch cartoon showing the development of the unconformity U1 and U2 and of the related seismic units in the post-collisional tectonic setting in northern offshore Taiwan. After reaching the culmination of orogenic activities in northern offshore Taiwan (~2.6 Ma; **Figure 8A**), the post-collisional magmatism was also about to begin. Afterward, the mountain range began to collapse, leaving the unconformity U1 at 1.6 Ma and SU II (**Figure 8B**). The initial subsidence was fault-controlled and was likely to be dominated by the rotation of the fault blocks, reflecting a mechanical stretching typical of early stages of rifting basin development [36, 37, 38]. The fault-controlled subsidence and volcanic activities may cease at late Quaternary (~0.2 Mal **Figure 8C**), resulting in a change in regional subsidence rate, a horizontally distributed unconformity U2, and a relatively even-thick SU III. In this model, the lowermost seismic unit SU I is a pre-Late Miocene sequence, representing the main body of collapsed imbricated thrusts. The middle seismic unit SU II caused by probably more rapid, fault-controlled subsidence indicates the depositional sequence of early stage mountain collapse. The uppermost seismic unit SU III, on the other hand, resulted from slower, probably thermal-control subsidence and indicates the depositional sequence of late stage of mountain collapse.

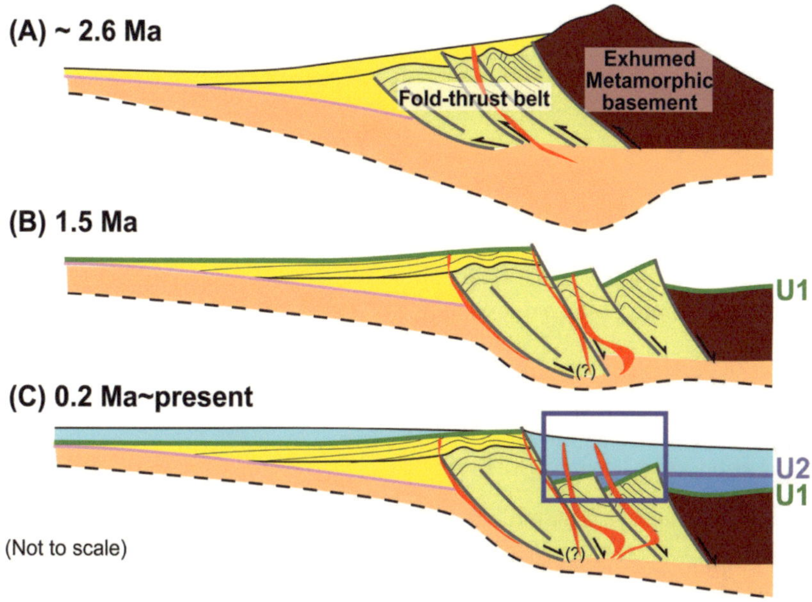

Figure 8. A schematic cartoon showing the development of the U1 and U2, with the tectonic scenario of post-collisional in northern Taiwan. Blue rectangle indicates the sites referring to our study area. We interpret the U1 and U2 indicate the onset and cessation of fault-controlled subsidence at 1.5 Ma and 0.2 Ma, respectively.

Also in this model, the volcanic magmatism of the NTVZ may be reducing since unconformity U2 appears to indicate the cessation of active extension. However, previous geophysical observations have shown that in addition to siliciclastic sediments, andesitic intrusions were also found at several places offshore northern Taiwan [39]. Recent observations about volcanic activities suggest that there are even younger, Holocene volcanic events in the NTVZ area, and hydrothermal activities in the Tatun Volcanic Group area are still very common at present [40, 41]. New seismological evidence clearly indicates that a deep magma reservoir is beneath Taipei [42]. Also, increasing active submarine volcano, gas plumes, and topography lineaments have been identified in the offshore area [43–45]. To sum up, investigation on whether the volcanic activities are still potential threatening geohazards, along with the geophysical and geochemical monitoring on present-day activities, remains important. We wish that the better understanding of the unconformities formation and relevant stratigraphic architectures will provide a good insight into Late Pleistocene-Holocene volcano-tectonic evolution of the NTVZ.

Acknowledgements

We are grateful to all the crew members of the R/V *Ocean Researcher II* who participated in the MG&G investigation cruise OR2-2047, and to science party members of the National Taiwan University who helped with marine geophysical data collection. We wish to thank Prof. Jui-Lin Chang, Dr. Shyh-Chin Lan of the GeoResource Research Center, National Cheng Kung University, Mr. Jian-Ming Chen of the Chinese Petroleum Company, Taiwan, and Prof. Jin-Oh

Park of the Atmosphere and Ocean Research Institute for their encouragement. We thank the Bureau of Mine, Ministry of Economic Affair and Ministry of Science and Technology, Taiwan for the financial support.

Author details

Jih-Hsin Chang[1,2]*, Eason Yi-Cheng Yang[1], Ho-Han Hsu[1,3], Chih-Chieh Su[1], Char-Shine Liu[1], Shye-Donq Chiu[1], Yu-Fang Ma[4], Yuan-Wei Li[5], Yen-Chun Lin[6] and Jen-Sen Shen[7]

*Address all correspondence to: changjihhsin@gmail.com

1 Institute of Oceanography, National Taiwan University, Taipei, Taiwan

2 Atmosphere and Ocean Research Institute, The University of Tokyo, Chiba, Japan

3 Institute of Earth Sciences, Academic Sinica, Taipei, Taiwan

4 Marine Exploration Instrumental Center, National Taiwan University, Taipei, Taiwan

5 Offshore Exploration & Production Division, Exploration & Production Business Division, Chinese Petroleum Company, Taipei, Taiwan

6 GeoResource Research Center, National Cheng Kung University, Tainan, Taiwan

7 Bureau of Mines, Ministry of Economic Affair, Taipei, Taiwan

References

[1] Chen TT, Watkins JS. Structure and stratigraphy of South Pengchiahsu Basin, northern offshore Taiwan. Petroleum Geology of Taiwan. 1994;**29**:127-170

[2] Cukur D, Horozal S, Kim DC, Han HC. Seismic stratigraphy and structural analysis of the northern East China Sea Shelf Basin interpreted from multi-channel seismic reflection data and cross-section restoration. Marine and Petroleum Geology. 2011;**28**:1003-1022. DOI: 10.1016/j.marpetgeo.2011.01.002

[3] Huang ST, Ting HH, Chen RC, Chi WR, Hu CC, Shen HC. Basinal framework and tectonic evolution of offshore northern Taiwan. Petroleum Geology of Taiwan. 1992;**27**:47-72

[4] Sibuet JC, Hsu SK. Geodynamics of the Taiwan arc-arc collision. Tectonophysics. 1997; **274**:221-251

[5] Lin JY, Sibuet JC, Hsu SK. Distribution of the East China Sea continental shelf basins and depths of magnetic sources. Earth Planets Space. 2005;**57**:1063-1072

[6] Teng LS. Extensional collapse of the northern Taiwan mountain belt. Geology. 1996; **24**(10):949-952

[7] Hsiao LY. Late Cenozoic structures off northeastern Taiwan [thesis]. National Taiwan University; 1997 62 p

[8] Hsiao LY, Lin KA, Huang ST, Teng LS. Structural characteristics of the southern Taiwan-Sinzi folded zone. Petroluem Geology of Taiwan. 1998;**32**:133-153

[9] Kong F, Lawver LA, Lee TY. Evolution of the southern Taiwan-Sinzi folded zone and opening of the southern Okinawa trough. Journal of Asian Earth Sciences. 2000;**18**:325-341

[10] Xu J, Ben-Avraham Z, Kelty T, Yu HS. Origin of marginal basins of the NW Pacific and their plate tectonic reconstructions. Earth-Science Reviews. 2014;**130**:154-196. DOI: dx.doi.org/10.1016/j.earscirev.2013.10.002

[11] Sun SC. The Cenozoic tectonic evolution of offshore Taiwan. Energy. 1985;**10**(3/4):421-432

[12] Lee TY, Hsu YY, Tang CH. Sequence stratigraphy and depositional cycles in the Tungyintao Basin, offshore northern Taiwan. Petroluem Geology of Taiwan. 1996;**30**:1-30

[13] Sibuet JC, Letouzey J, Barbier F, Charvet J, Foucher JP, Hlide TWC, Kimura M, Chiao LY, Marsset B, Muller C, Stephan JF. Back arc extension in the Okinawa Trough. Journal of Geophysical Research. 1987;**92**(B13):14,041-14,063

[14] Sibuet JC, Deffontaines B, Hsu SK, Thareau N, Le Formal JP, Liu CS. ACT party. Okinawa trough Backarc Basin: Early tectonic and magmatic evolution. Journal of Geophysical Research. 1998;**103**(B12):30245-30267

[15] Park JO, Tokuyama H, Shiohara M, Suyehiro K, Taira A. Seismic record of tectonic evolution and backarc rifting in the southern Ryukyu island arc system. Tectonophysics. 1998;**294**:21-37

[16] Hsu SK, Sibuet JC, Shyu CT. Magnetic inversion in the East China Sea and Okinawa trough: Tectonic implications. Tectonophysics. 2001;**333**:111-122

[17] Ho CS. An Introduction to the Geology of Taiwan: Explanatory Text of the Geological Map of Taiwan, Central Geological Survey, Ministry of Economic Affairs, Taiwan. 2nd ed 1988 192 p

[18] Teng LS. Geotectonic evolution of late Cenozoic arc-continent collision in Taiwan. Tectonophysics. 1990;**183**:57-76

[19] Teng LS. Geotectonic evolution of Tertiary contienental margin basins od Taiwan. Petroluem Geology of Taiwan. 1992;**27**:1-19

[20] Lin AT, Yao B, Hsu SK, Liu CS, Huang CY. Tectonic features of the incipient arc-continent collision zone of Taiwan: Implications for seismicity. Tectonophysics. 2009;**479**:28-42. DOI: 10.1016/j.tecto.2008.11.004

[21] Huang ST, Yang KM, Huang JH, Wu JC, Ting HH, Mei WW, Hsu SK, Lee M. Deformation front development at the northeast margin of the Tainan Basin, Tainan-Kaohsiung area, Taiwan. Marine Geophysical Researches. 2004;**25**:139-156. DOI: 10.1007/s11001-005-0739-z

[22] Lin CH. Continental subduction and active crustal exhumation in Central Taiwan. In: Workshop on Tectonic Evolution of Taiwan, Mar 26, 2010, Academia Sinica, Institute of Earth Sciences. 2010

[23] Teng LS, Lee CT, Peng CH, Chen WF, Chu CJ. Origin and geological evolution of the Taipei Basin, northern Taiwan. Western Pacific Earth Science. 2001;**1**(2):115-142

[24] Emery KO, Niino H. Stratigraphy and petroleum prospects of Korea Strait and the East China Sea. Coordinating Committee for Geoscience Programmes in East and Southeast Asia Technical Bulletin. 1968;**1**:13-27

[25] Emery KO, Hayashi Y, Hlide TWC, Kobayashi K, COO JH, Meng CY, Niino H, Osterhgen JH, Reynolds LM, Wageman JM, Wang CS, Yang SJ. Geological structures and some water characteristics of the East China Sea and the Yellow Sea. Committee for Coordination of Joint Prospecting Technical Bulletin. 1969;**2**:3-43

[26] Wageman JM, Hlide TWC, Emery KO. Structural framework of East China Sea and Yellow Sea. The American Association of Petroleum Geologists Bulletin. 1970;**54**(9):1611-1643

[27] Letouzey J, Kimura M. The Okinawa Trough: Genesis of a back-arc basin developing along a continental margin. Tectonophysics. 1986;**125**:209-230

[28] Zhu W, Mi L. Atlas of oil and gas basins. China Sea: Petroleum Industry Press; 2010 316 p

[29] Gungor A, Lee GH, Kim HJ, Han HC, Kang MH, Kim J, Sunwoo D. Structural characteristics of the northern Okinawa trough and adjacent areas from regional seismic reflection data: Geologic and tectonic implications. Tectonophysics. 2012;**522-523**:198-207. DOI: 10.1016/j.tecto.2011.11.027

[30] Wang KL, Chung SL, O'reilly SY, Sun SS, Shinjo R, Chen CH. Geochemical constraints for the genesis of post-collisional magmatism and the geodynamic evolution of the northern Taiwan Region. Journal of Petrology. 2004;**45**(5):975-1011. DOI: 10.1093/petrology/egh001

[31] Belousov A, Belousova M, Chen CH, Zellmer GF. Deposits, character and timing of recent eruptions and gravitational collapses in Tatun Volcanic Group, northern Taiwan: Hazard-related issues. Journal of Volcanology and Geothermal Research. 2010;**191**:205-221. DOI: 10.1016/j.jvolgeores.2010.02.001

[32] Liu CS, Liu SY, Lallemand SE, Lundberg N, Reed DL. Digital elevation model offshore Taiwan and its tectonic implication. Terrestrial, Atomspheric and Oceanic Science. 1998;**9**(4):705-738

[33] Wessel P, Smith WHF, Scharroo R, Luis JF, Wobbe F. Generic mapping tools: Improved version released. EOS Transactions, AGU. 2013;**94**:409-410. DOI: 10.1002/2013EO450001

[34] Vail PR, Mitchum RM. Seismic stratigraphy and global changes of sea level, part 1: Overview. In: Payton CE, editor. Seismic Stratigraphy-applications to Hydrocarbon Exploration. American Association of Petroleum Geologist U.S.A; 1977. p. 51-52

[35] Mitchum RM, Vail PR, Thompson S. Seimic stratigraphy and global changes of sea level, part 2: The depositional sequence as a basic unit for stratigraphic analysis. In: Payton CE, editor. Seismic Stratigraphy-applications to Hydrocarbon Exploration. American Association of Petroleum Geologist Memoir 26, Tulsa, Oklahoma, U.S.A. 1977. p. 53-62

[36] Prosser S, Rift-related linked depositional systems and their seismic expression. In: Willams GD and Dobb A, editor. Tectonics and seismic sequence stratigraphy. Geological Society London Special Publication. 1993;**71**:35-66

[37] Klein G, Hsui AT. Origin of cratonic basins. Geology. 1987;**15**:1094-1098

[38] Lee TY, Tang CH, Ting JS, Hsu YY. Sequence stratigraphy of the Tainan Basin, offshore southwestern Taiwan. Prteoluem Geology of Taiwan. 1993;**28**:119-158

[39] Yen TP. A geologic consideration on the onland and o shore of northern Taiwan. Proceedings of the Geological Society of China. 1980;**23**:37-45

[40] Ohba T, Sawa T, Taira N, Yang TF, Lee HF, Lan TF, Ohwada M, Morikawa N, Kazahaya K. Magmatic fluids of Tatun volcanic group, Taiwan. Applied Geochemistry. 2010;**25**:513-523. DOI: 10.1016/j.apgeochem.2010.01.009

[41] Lee HF, Yang TF, Lan TF, Chen CH, Song SR, Tsao SJ. Temporal variations of gas compositions of fumaroles in the Tatun Volcano Group, northern Taiwan. Journal of Volcanology and Geothermal Research. 2008;**178**:624-635. DOI: 10.1016/j.jvolgeores.2008.06.005

[42] Lin CH. Evidence for a magma reservoir beneath the Taipei metropolis of Taiwan from both S-wave shadows and P-wave delays. Scientific Report. 2016;**6**(39500):1-9. DOI: 10.1038/srep39500(2016)

[43] Tsai CH, Hsu SK, Lin SS, Yang TF, Wang SY, Doo WB, Lee HF, Lan TF, Huang JC, Liang CW. The Keelung Submarine Volcano in the near-shore area of northern Taiwan and its tectonic implication. Journal of Asian Earth Sciences. 2017;**135**:320-326. DOI 10.1016/j.jseaes.2016.12.041

[44] Hung T, Song G. The tracing of o shore extension of the Jinshan active fault of North Taiwan. In: American Geophysical Union Fall Meeting 2012. San Francisco: AGU; 2012

[45] Hung TI. Geomorphologic study offshore the Jingshan area [thesis]. National Taiwan University; 2013 54 p

Section 3

Integrated Stratigraphy

Chapter 5

Stratigraphy of Jurassic Sediments of the Southern Siberian Platform (Russia) Studied through Lithologic and Paleobotanical Data

Andrey Olegovich Frolov,
Nikolay Ivanovich Akulov and
Irina Mikhailovna Mashchuk

Additional information is available at the end of the chapter

http://dx.doi.org/10.5772/intechopen.69572

Abstract

This paper presents the results of comprehensive lithologic and paleobotanical study of Jurassic sediments of the Irkutsk Coal Basin revealed in outcrops and sections within operating coal deposits. The lithologic characteristics of the main stratigraphic units of the Irkutsk Basin: Cheremkhovskaya, Prisayanskaya, and Kudinskaya Formations are given. Two uneven-aged fossil plant assemblages: Cheremkhovo for middle and upper subformations of Cheremkhovskaya Formation and Prisayan for Prisayanskaya and Kudinskaya Formations have been identified. *Equisetites lateralis* (Phill.) Phill., *E. asiaticus* Pryn., *Cladophlebis haiburnensis* (L. et H.) Sew., *Sphenobaiera czekanowskiana* (Heer) Flor., *S. vigentis* Kiritch. et Bat., *Czekanowskia baikalica* Kiritch. et Samyl., and *Cz. rigida* Heer species are typical of the Cheremkhovo assemblage. The age of sediments including the Cheremkhovo assemblage is the end of Early Jurassic (conventionally, Toarcian). *Coniopteris maakiana* (Heer) Pryn. emend. Kiritch. et Trav., *C. murrayana* (Brongn.) Brongn., *C. spectabilis* Brick., *Cladophlebis nebbensis* (Brongn.) Nath., *Raphaelia diamensis* Sew., *R. tapkensis* (Heer) Pryn. emend Kost., *Phoenicopsis angustifolia* Heer, *Ph. cognata* Kiritch., and *Ph. irkutensis* Dolud. et Rasskaz species are characteristic of the Prisayan assemblage. Prisayan assemblage sediments are dated at the beginning of Middle Jurassic (conventionally, Aalenian). The stratigraphic correlation of Jurassic sediments of the Irkutsk Coal Basin with the sedimentary basins of Western Siberia has been carried out based on paleobotanical data.

Keywords: Jurassic sediments, stratigraphy, fossil and flora assemblages, Irkutsk Basin, Siberian platform

1. Introduction

Lower and middle Jurassic continental sediments of Siberia are abundant. They are exposed along river valleys and gorges, within quarries and mine workings of the Kuznetsk, Kansk, Irkutsk, and other Coal Basins. Lower and Middle Jurassic sediments are rich in plant remains, which are important in stratigraphy and correlation of continental complex of Jurassic sediments where large coal deposits are concentrated. The Irkutsk Basin situated within the southern Siberian platform is one of them (**Figure 1**). Three structural and facies zones: Prisayan piedmont trough, Platform limb, and Angara-Koty intermountain area are clearly distinguished within it [2]. According to the regional stratigraphic scheme, the Jurassic sediments of the Platform limb and Prisayan piedmont trough are subdivided into three formations: Cheremkhovskaya, Prisayanskaya, and Kudinskaya. Cheremkhovskaya Formation

Figure 1. Layout of the studied key sections. 1—the left bank of the Bol'shaya Belaya river at 200 m below the railway bridge in the Taiturka settlement (GPS: N52°86.242'; E103°49.371'); 2—the right bank of the Zalari river, on the opposite side of the Zalari settlement (GPS: N53°55.924'; E102°55.199'); 3—the Cheremkhovo coal deposit (GPS: N53°20.329'; E103°11.985'); 4—the Glinki coal deposit (GPS: N53°86.684'; E102°26.406'); 5—the Mugun coal deposit (GPS: N54°43.050'; E100°18.245'); 6—the right bank of the Angara river, the Ust'-Balei creek (GPS: N52°62.771'; E103°96.128'); 7—the right bank of the Angara river, the Tolstyi Cape (GPS: N52°63.714'; E103°93.978'); 8—the right bank of the Angara river, on the opposite side of the Tel'ma settlement (GPS: N52°70.676'; E103°77.649'); 9—the right bank of the Irkut river, Kaiskaya Gora (GPS: N52°28.331'; E104°23.019'); 10—Sinyushina Gora (GPS: N52°15.888'; E104°11.188'); 11—the Olkha river, the Smolenshchina settlement (GPS: N52°15.006'; E104°9.305'); 12—the Olkha river, Olkhinskaya Gora area; 13—the left bank of the Irkutsk water reservoir, on the opposite side of the Stroitel' dacha (GPS: N52°11.195'; E104°23.459'); 14—the right bank of the Angara river, the Topka creek valley (GPS: N52°21.289'; E104°17.282'); 15—the left bank of the Irkut river, at 2 km above the Pionersk settlement (GPS: N52°23.734'; E103°99.541'); 16—the right bank of the Kuda river, 2 km to the north of the Zherdovka settlement (GPS: N52°66.490'; E104°57.222').

is composed of three subformations: lower, middle, and upper and Prisayanskaya includes lower and upper subformations. The stratigraphic sequences of these formations and their lithologic characteristics were confirmed by lithostratigraphic data in different boreholes and natural sections [1, 28].

Since the second half of the nineteenth century, the Early-Middle Jurassic flora of the Irkutsk Basin were repeatedly studied in detail. Data on Jurassic flora of the Basin are cited in various works [14–16, 22, 27, 29]. The results of the studies on some groups of fossil plants are published in several papers [4–6, 8–12, 18–20, 25]. Despite the large number of publications, there is no unified view on stratigraphy of Jurassic sediments of the Irkutsk Basin. Against the background of well-studied Jurassic floras of the Kuznetsk [17], the Kansk Basins [24], and Western Siberia [21], Jurassic flora is still poorly studied. Paleobotanical characteristic of formations needs clarification, and stratigraphic importance of some species of fossil plants requires revision. For example, species *Phlebopteris polypodioides* Brongn and *Clathropteris obovata* Oishi indicated in regional stratigraphic scheme [2] are rare, and some representatives of genus *Coniopteris* are abolished [*C. clavipes* (Heer) Pryn., *C. trautscholdii* (Heer) Pryn.] [18], consequently, they cannot be used in stratigraphy.

The aim of this work is to suggest a solution for the above-listed problems by the implementation of comprehensive lithologic and paleobotanical study of key sections of Jurassic sediments within present-day active coal deposits rich in paleobotanical material.

2. Methods

Starting in 2008 and in 2016, the authors carried out quite a number of field works in order to explore lithologic and paleobotanical peculiarities in all active coal deposits within the Irkutsk Basin (see **Figure 1**). Opencasts of three coal-producing quarries: Cheremkhovo, Mugun, and Glinki were studied in detail. In addition, 13 exposures which are stratotypes of regional stratigraphic units were studied. The leaves of Ginkgoales and Leptostrobales were investigated by the epidermal-cuticular method. Leaf cuticles were macerated by the standard technique in Schultze mixture. Microslides were examined using the Olympus BX41TF light microscope and electronic scanning microscope Philips' SEM 525-M. About one hundred pieces of ore and more than 300 microslides with unfolded pattern of leaves' compressions were studied.

3. Lithostratigraphy of the formations of the Irkutsk Basin

The Cheremkhovskaya Formation was first identified by Korovin [23] in mines within the Cheremkhovo settlement vicinity. It represents a complete megarhythm and is divided into three subformations: lower, middle, and upper ones with total thickness up to 380 m.

Lower Cheremkhovskaya (Zalarinskaya) Subformation occurs with discordance on Cambrian limestones of Angarskaya Group. The subformation base is composed of clastic deposits—conglomerates, gravelites, and coarse-grained sandstones which are overlapped by siltstones, mudstones with thin coal bands.

On the right bank of the Zalari river, on the opposite side of the Zalari settlement is situated the stratotype of Lower Cheremkhovskaya Subformation (bottom-up):

		Thickness, m
1.	Conglomerates interstratifying with coarse-grained sandstones	0.65
2.	Sandstones, coarse-grained, massive	1.34
3.	Conglomerates, fine-pebbled	2.00
4.	Sandstones, white, coarse-grained, massive	0.30
5.	Conglomerates, fine-pebbled	0.80
Disappearance of outcrop		3.00
6.	Sandstones, gray, medium-fine-grained with horizontal bedding and abundant plant detrital matter	0.70
Disappearance of outcrop		3.50
7.	Sandstones, pale gray, coarse-grained, massive	0.60
8.	Sandstones, gray, medium-grained with horizontal bedding	0.60
Disappearance of outcrop		1.39
9.	Sandstones, red, medium-fine-grained with horizontal bedding	1.10
10.	Alternation of red, burnt rocks presented by sandstones and siltstones with horizontal bedding and impressions of *Cladophlebis haiburnensis* (L. et H.) Sew., *Sphenobaiera ex gr. czekanowskiana* (Heer) Flor., *Czekanowskia ex gr. rigida* Heer, *Podozamites eichwaldii* Pryn. var. *minor* Pryn. *Ixostrobus grandis* Tesl.	3.54
11.	Speckled sandstones, coarse-medium-grained with horizontal bedding	0.75
Developed thickness		13.12

Clastic deposits of the lower subformation are traced by us in some natural outcrops along the ranks of the Bol'shaya Belaya river.

Industrially coal-bearing **Middle Cheremkhovskaya Subformation** is a natural sedimentary continuation of the lower subformation, and it is related to the latter by gradual transitions. This subformation is characterized by abundance in section of siltstones, mudstones, and coal seams with thickness up to 10–15 m. The thickness was cumulated during the epoch of maximum regional coal storage within the Basin and is abundant all over. Lower boundary line is drawn in bottom of siltstones and sandstones underlying the first coal-bearing horizon.

		Thickness, m
Middle Cheremkhovskaya Subformation		
1.	Siltstones, gray	0.60
2.	Coal hard	0.50
3.	Mudstones, coaly with horizontal bedding	0.20
4.	Coal	0.30
5.	Mudstones, coaly with horizontal bedding	0.20
6.	Coal	0.50
7.	Mudstones, coaly with horizontal bedding	0.25
8.	Coal	0.80
9.	Sandstones, gray, medium-fine-grained with horizontal bedding, contain plant remains of *Czekanowskia* ex gr. *rigida* Heer	1.00
10.	Siltstones, gray	0.40
11.	Sandstones, gray, medium-fine-grained	1.45
12.	Siltstones, gray, with horizontal bedding	0.30
13.	Sandstones, gray, medium-fine-grained	0.50
14.	Mudstones, coaly, with horizontal bedding	0.20
15.	Sandstones, gray, fine-grained with horizontal bedding	2.00
16.	Siltstones, gray, with horizontal bedding, contain plant remains of *Equisetites lateralis* (Phill) Phill., *Cladophlebis haiburnensis* (L. et H.) Sew., *Cl. williamsonii* Brongn., *Raphaelia diamensis* Sew., *Sphenobaiera czekanowskiana* (Heer) Flor., *S. vigentis* Kiritch. et Bat., *Pseudotorellia paradoxa* Dolud., *Czekanowskia rigida* Heer, *Cz. baikalica* Kiritch. et Samyl., *Pityophyllum ex gr. nordenskioldii* (Heer) Nath., *Carpolithes cinctus* Nath., *Ixostrobus heeri* Pryn., and *Ix. grandis* Tesl.	1.20
17.	Coal	1.00
18.	Mudstones, coaly with horizontal bedding, contain fossil plant remains of *Cladophlebis* sp., *Czekanowskia baikalica* Kiritch. et Samyl., and *Pityophyllum* ex gr. *nordenskioldii* (Heer) Nath	0.20
19.	Coal	0.78
20.	Mudstones, coaly with horizontal bedding	0.10
21.	Coal	0.45
22.	Mudstones, coaly with horizontal bedding	0.14
23.	Coal	0.40
24.	Siltstones, gray, rich in plant remains of *Lycopodites sp.*, *Czekanowskia* ex gr. *rigida* Heer, *Phoenicopsis* ex gr. *angustifolia* Heer, *Pityophyllum* ex gr. *nordenskioldii* (Heer) Nath., and *Schizolepis follinii* Nath.	0.64
25.	Coal	0.20

		Thickness, m
26.	Siltstones, gray, with burst plant detritus	0.30
27.	Sandstones, yellow, medium-fine-grained with impressions of *Czekanowskia* ex gr. *rigida* Heer and *Pityophyllum* ex gr. *nordenskioldii* (Heer) Nath.	2.00
28.	Coal	0.20
29.	Sandstones, yellowish-gray, coarse-grained	1.65
30.	Sandstones, gray, medium-fine-grained with leaf impressions of *Czekanowskia* ex gr. *rigida* Heer and coalified plant detritus	0.80
31.	Coal	0.20
32.	Sandstones, yellowish-gray, medium-fine-grained, poorly cemented, flat-bedded with abundant plant detritus	0.80
33.	Coal	0.20
Upper Cheremkhovskaya Subformation		
34.	Sandstones, yellow, medium-grained	4.50
35.	Conglomerates, fine-pebbly	0.30
36.	Sandstones, yellow, medium-grained polymictic	4.60
Thickness of the opencast developed part		29.47

Section uncovering within the Cheremkhovo coal deposit situated in the central part of the Irkutsk Basin is a stratotype of Middle Cheremkhovskaya Subformation (bottom-up) (**Figure 2**):

Middle Cheremkhovskaya Subformation is opened in coal opencasts of the Glinki and Mugun deposits. The Glinki coal-bearing deposits occur on erosional contacts with a weathered surface of the Upper Cambrian clays (**Figure 2**). Opened thickness of Jurassic deposits is 18.4 m. Middle Cheremkhovskaya Subformation composed of gray, flat bedded, fine-grained sandstones, siltstones, and mudstones with two coal seams of industrial thickness (2–5.8 m) is exposed in the lower part of the opencast. We revealed the following plant remains: *Equisetites asiaticus* Pryn., *Equisetites* cf. *lateralis* (Phill.) Phill., *Hausmannia crenata* (Nath.) Maell., *Cladophlebis williamsonii* Brongn., *Cl. haiburnensis* (L et H) Sew., *Ginkgo concinna* Heer, *Sphenobaiera czekanowskiana* (Heer) Florin, *Pseudotorellia* cf. *paradoxa* Dolud., *Czekanowskia baikalica* Kiritch. et Samyl., *Cz. rigida* Heer, *Phoenicopsis* ex gr. *angustifolia* Heer, *Leptostrobus laxiflora* Heer, *Podozamites* cf. *lanceolatus* (L. et H.) Schimp., *P.* cf. *eichwaldii* Pryn. var. *major* Pryn., *Pityophyllum* ex gr. *nordenskioldii* (Heer) Nath., *Schizolepis* cf. *moelleri* Sew., *Carpolithes minor* Pryn., and *Ixostrobus heeri* Pryn. in siltstones and mudstones of industrially coal-bearing part of the opencast.

The Mugun lignite deposit is situated in the northwestern part of the Irkutsk Basin, at 40 km to the south of the Tulun city. The deposit has an irregular shape and a simple geological structure. Lower band of the deposit belongs to Middle Cheremkhovskaya Subformation. It is composed of siltstones, mudstones, and fine-grained sandstones alternating among themselves. Productive coal seams (thickness from 3.5 up to 10 m) are confined to this part of

http://dx.doi.org/10.5772/intechopen.69572

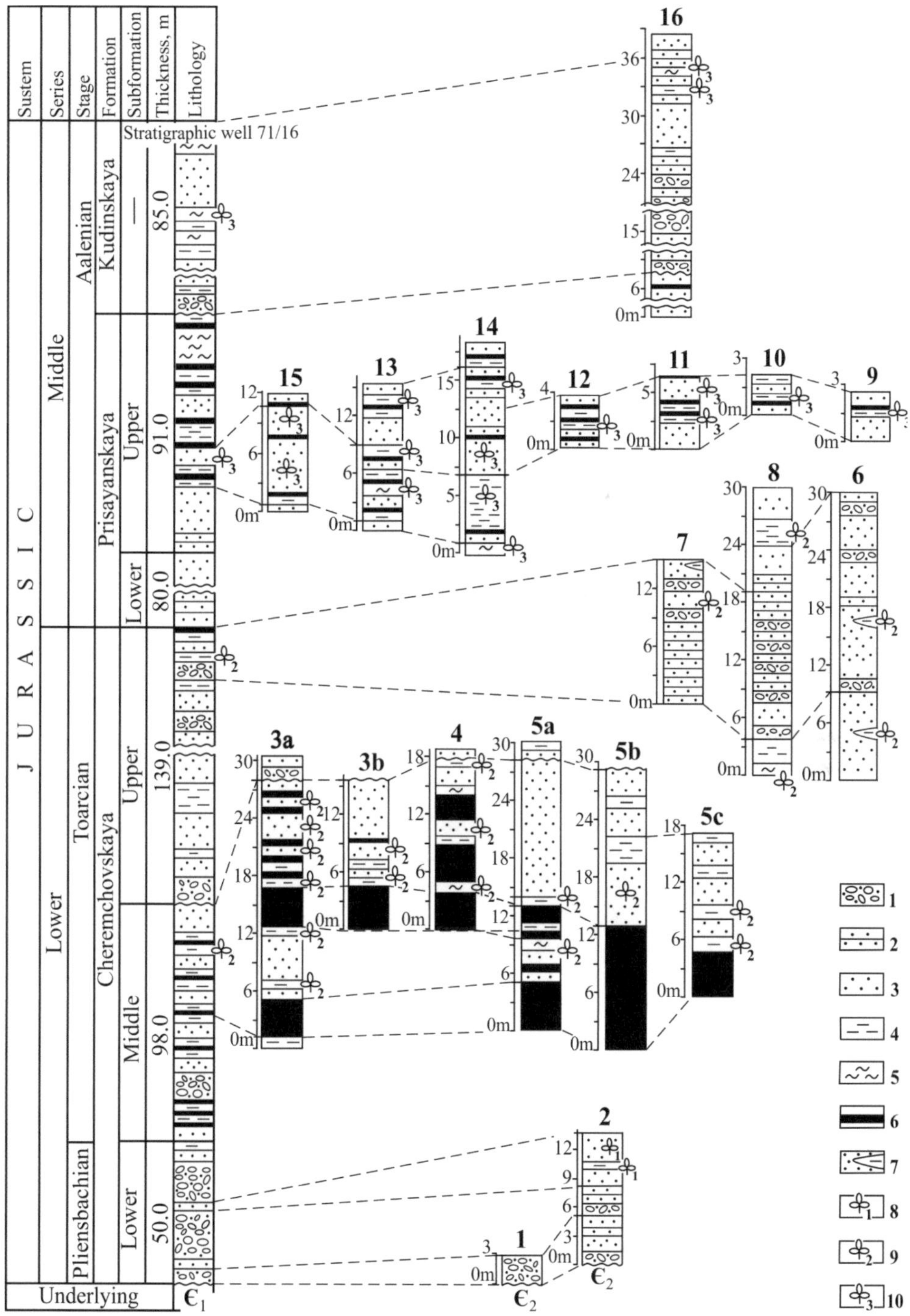

Figure 2. Correlation scheme of the key sections of Jurassic sediments of the Irkusk Basin. Numbering of sections is given in **Figure 1**. 1—conglomerates, 2—coarse-grained sandstones, 3—medium- and fine-grained sandstones, 4—siltstones, 5—mudstones, 6—coals, 7—lenses; plant remains: 8—with no stratigraphic importance, 9—typical of Cheremkhovo plant assemblage, 10—typical of Prisayan plant assemblage.

opencast. We distinguished the following plant remains: *Coniopteris* sp., *Lobifolia nana* A. Frol., *Czekanowskia rigida* Heer, *Pityophyllum* ex gr. *nordenskioldii* (Heer) Nath., and *Carpolithes heeri* Tur.-Ket. [10] in siltstones and mudstones of this band.

Upper Cheremkhovskaya (Ust'-Baleiskaya) Subformation overlaps in concordance with the Middle Cheremkhovskaya Subformation, and more than a half of the former is composed of fine- and medium-grained sandstones with lenses of siltstones and mudstones. Lower boundary line is drawn in the bottom of sandstones of channel facies [1] and it is observed within opencasts of developed coal deposits (Cheremkhovo, Glinki, and Mugun; **Figure 2**). Outcrop situated at 2.1 km below the Ust'-Baley settlement is a stratotype of the upper subformation. Gray, differently grained polymictic sandstones with horizontal cross, uni-, and diversidirectional lamination dominate in its lower part. Sandstones contain two lenses composed of siltstones and mudstones with horizontal and sometimes banded lamination. Lower lens is long about 70 m with thickness up to 2 m. Impressions of the following plants: *Lycopodites tenerrimus* Heer, *Czekanowskia rigida* Heer, *Elatides ovalis* Heer, *Ixostrobus* sp., *Carpolithes deplanatus* Pryn., *C. cinctus* Nath., and *Samaropsis rotundata* Heer are found on the surface of mudstone bedding.

Thickness of the upper lens reaches 5.8 m, and its length is about 120 m. Siltstones and mudstones of the lens contain numerous impressions of insects and fewer those of fishes and shells of limnetic myarians. Plant remains of *Lycopodites tenerrimus* Heer, *L. trichiatus* Pryn. emend. A. Frol. et Mash., *Phyllotheca sibirica* Heer, *Coniopteris murrayana* (Brongn.) Brongn., *Sphenobaiera czekanowskiana* (Heer) Florin, *Czekanowskia rigida* Heer, *Leptostrobus laxiflora* Heer, *Elatides ovalis* Heer, *Ixostrobus heeri* Pryn., and *Samaropsis rotundata* Heer are found on several plates of beddings.

Outcrops situated on the opposite side of the Tel'ma settlement (**Figure 2**) and the Tolstyi Cape have similar lithologic structure. The following plant remains: *Cladophlebis* cf. *haiburnensis* (L. et H.) Sew., *Cl. williamsonii* Brongn., *Sphenobaiera czekanowskiana* (Heer) Flor., *S. vigentis* Kiritch. et Bat., *Pseudotorellia longifolia* Dolud., *Czekanowskia rigida* Heer, *Cz. baikalica* Kiritch. et Samyl., *Leptostrobus laxiflora* Heer, *Pityophyllum* ex gr. *nordenskioldii* (Heer) Nath., *Schizolepis* sp., *Samaropsis rotundata* Heer, and *Carpolithes* cf. *minor* Pryn. are revealed in siltstone lens opened by these outcrops.

Prisayanskaya Formation is less abundant relative to the Cheremkhovskaya one which is overlapped with conformability by it.

Lower Prisayanskaya (Idanskaya) Subformation is predominantly confined to the southeastern part of the Basin, its thickness is of 50–70 m. Characteristic feature of its sections is the prevalence of coarse-grained and gritty sandstones as well as siltstones containing indeterminable plant remains of poor preservation. Conglomerates and gritstones occur on several areas in the base of subformation. Lower boundary line of subformation is drawn relative to band bottom of rhythmically alternating sandstones, gritstones, conglomerates, and siltstones. It is sufficiently conventional.

Upper Prisayanskaya (Sukhovskaya) Subformation is preserved from erosion only in Priirkut depression. This subformation is composed of small and fine-grained sandstones and siltstones, and coarse-grained varieties of sandstones are rare. The characteristic feature of this subformation is the significant enrichment of rocks in coaly matter and the presence of coal bands. Thickness of this subformation is about 50 m. Its lower boundary with the lower subformation is not always distinct. Within Priirkut depression we studied outcrops of this Upper Prisayanskaya Subformation situated along the banks of the Angara, Irkur, Kaya, and Topka rivers.

In the quarry situated on Kaiskaya Gora at 400 m above the Kaya river mouth in the Irkutsk city the Jurassic deposits of Upper Prisayanskaya Subformation are opened (bottom-up).

		Thickness, m
1.	Sandstones, yellowish-gray, fine- and medium-grained, contain plant remains of *Cladophlebis williamsonii* Brongn., *Phoenicopsis angustifolia* Heer, and *Pityophyllum* ex gr. *nordenskioldii* (Heer) Nath.	2.00
2.	Siltstones and silty sandstones, yellowish-gray, with impressions of plants: *Lobifolia lobifolia* (Phill.) Rasskaz. et E. Leb., *Coniopteris maakiana* (Heer) Pryn. emend. Kiritch. et Trav., *Cladophlebis williamsonii* Brongn., *Czekanowskia* ex gr. *rigida* Heer, *Phoenicopsis angustifolia* Heer, *Ph. Markovitchiae* Kiritch. et Schischk., *Ph. dentata* Pryn., Ph. irkutensis Dolud. et Rasskaz., and *Pityophyllum* ex gr. *nordenskioldii* (Heer) Nath.	0.20
3.	Coal	0.40
4.	Sandstones, yellowish-gray, fine-grained, contain plant impressions of *Coniopteris maakiana* (Heer) Pryn. emend. Kiritch. et Trav., C. cf. *sachsii* Tesl.	0.20
Opened thickness		2.80

The section of Upper Prisayanskaya Subformation is opened within Sinyushina Gora in the Irkutsk city. Gray, fine- and medium-grained sandstones dominate in the lower part of the section. The section's upper part is composed of alternating medium-grained sandstones and siltstones, only one thin (0.07 m) coal band is noted. *Coniopteris murrayana* (Brongn.) Brongn., *Cladophlebis nebbensis* (Brongn.) Nath., *Cl. williamsonii* Brongn., *Czekanowskia* ex gr. *rigida* Heer, *Phoenicopsis angustifolia* Heer, *Ph. samylinae* Kiritch. et Moskv., *Ph. irkutensis* Dolud. et Rasskaz., and *Pityophyllum* ex gr. *nordenskioldii* (Heer) Nath are revealed among plant remains.

The upper part of Upper Prisayanskaya Subformation of 5.4 m thickness and 400 m length is opened in roadside quarry near the Smolenshchina settlement. The section is of rhythmic structure. In this section, quartz-feldspathic coarse- and medium-grained sandstones alternate upward with fine-grained sandstones, siltstones, and mudstones with thin (0.1–0.3 m) coal bands. *Hepaticites arcuatus* (L. et H.) Harris, *Coniopteris murrayana* (Brongn.) Brongn., *C. maakiana* (Heer) Pryn. emend. Kiritch. et Trav., *C. spectabilis* Brick, *Cladophlebis williamsonii* Brongn., *Cl. haiburnensis* (L. et H.) Brongn., *Cl.* cf. *nebbensis* (Brongn.) Nath., *Raphaelia diamensis* Sew., *Sphenobaiera* ex gr. *czekanowskiana* (Heer) Flor., *Ginkgo* ex gr. *sibirica* Heer, *Czekanowskia* ex gr. *rigida* Heer, *Phoenicopsis angustifolia* Heer, *Ph.* cf. *mogutchevae* Kiritch. et Trav., *Ph. cognata* Kiritch., *Pityophyllum* ex gr. *nordenskioldii* (Heer) Nath., and *Stenorachis* (?) *clavata* Nath. are identified among plant remains.

One of the key sections of the Upper Prisayanskaya Subformation is outcrops situated on the left bank of the Irkutsk water reservoir on the opposite side of the Stroitel' settlement (**Figure 2**). The section has a rhythmic structure. Bases of rhythms are presented by quartz fieldspathic coarse- and medium-grained sandstones alternating upward the section with fine-grained sandstones, siltstones, and mudstones with thin (0.1–0.3 m) coal bands. Near water's edge are exposed the following rocks:

		Thickness, m
1.	Sandstones, gray medium-grained, horizontally bedded micaceous	0.14
2.	Sandstones, silty, gray, plant detritus is exposed, plant detritus is found in the upper part of the layer	0.46
3.	Hard coal	0.12
4.	Mudstones, gray	0.04
5.	Hard coal	0.16
6.	Mudstones, coaly, horizontally bedded with abundant plant detritus	0.13
7.	Hard coal	0.07
8.	Mudstones, coaly, horizontally bedded, contain abundant plant detritus of *Coniopteris maakiana* (Heer) Pryn. emend. Kirich. et Trav., *Cladophlebis* cf. *williamsonii* Brongn., *Raphaelia diamensis* Sew., *R. tapkensis* (Heer) Pryn. emend. Kost., *Czekanowskia* ex gr. *rigida* Heer, and *Phoenicopsis* ex gr. *angustifolia* Heer	0.11
9.	Sandstones, gray, medium-grained with plant detritus	0.07
10.	Sandstones, gray medium-grained with distinct cross bedding and impressions of trunks. Leaf mats of *Phoenicopsis* ex gr. *angustifolia* Heer are found at 0.4 m from the bottom layer	1.10
11.	Sandstones, gray, horizontally bedded with plant detritus	0.10
12.	Mudstones, gray, horizontally bedded. Impressions of *Coniopteris maakiana* (Heer) Pryn. emend. Kirich. et Trav., *Cladophlebis nebbensis* (Brongn.) Nath., *Cl. williamsonii* Brongn., *Cl. haiburnensisi* (L. et H.) Sew., *Raphaelia diamensis* Sew., *R. tapkensis* (Heer) Pryn. emend. Kost., *Sphenobaiera* ex gr. *czekanowskiana* (Heer) Florin, *Pseudotorellia* cf. *ensiformis* (Heer) Dolud., *P.* cf. *paradoxa* Dolud., *Czekanowskia* ex gr. *rigida* Heer, *Phoenicopsis angustifolia* Heer, *Elatocladus manchuricus* (Yok.) Yabe, and *Pityophyllum* ex gr. *nordenskioldii* (Heer) Nath. are found in the lower part of the layer	0.10
13.	Sandstones, gray, medium-grained with cross bedding and plant detritus	0.90
14.	Hard coal	0.06
15.	Mudstones with plant remains, poorly preserved	0.03
16.	Coal	0.04
17.	Mudstones, gray, with impressions of fossil plant remains: *Coniopteris maakiana* (Heer) Pryn. emend. Kiritch. et Trav., *C. murrayana* (Brongn.) Brongn., *Cladophlebis haiburnensis* (L. et H.) Sew., *Cl. williamsonii* Brongn., *Phoenicopsis* ex gr. *angustifolia* Heer, and *Pityophyllum* ex gr. *nordenskioldii* (Heer) Nath.	0.55
18.	Sandstones, medium-grained, massive	0.60

		Thickness, m
19.	Sandstones, horizontally bedded with impressions of *Lycopodites baicalensis* A. Frol., *Cladophlebis* cf. *nebbensis* (Brongn.) Nath., *Phoenicopsis* ex gr. *angustifolia* Heer, and *Pityophyllum* ex gr. *nordenskioldii* (Heer) Nath.	0.25
20.	Coal	0.02
21.	Sandstones, gray, fine-grained with thin slab parting	0.23
22.	Sandstones, gray, coarse-grained, cross-bedded with trunks of trees	4.00
23.	Sandstones, gray, medium-grained with two coal bands (0.02 and 0.03 m)	0.35
24.	Siltstones, gray, with plant remains: *Cladophlebis nebbensis* (Brongn.) Nath., *Cl. haiburnensis* (L. et H.) Sew., *Cl. williamsonii* Brongn., *Pityophyllum* ex gr. *nordenskioldii* (Heer) Nath., *Carpolithes cinctus* Nath.	0.10
25.	Sandstones, gray, fine-grained, cross-bedded, with impressions of *Pityophyllum* ex gr. *nordenskioldii* (Heer) Nath., poorly preserved	2.00
26.	Mudstones, coaly, with coal band (0.03 m)	0.23
27.	Mudstones, gray, with impressions of ancient plants: *Lycopodites baikalensis* A. Frol., *Hausmannia crenata* (Nath.) Maell., *Raphaelia* cf. *tapkensis* (Heer) Pryn. emend. Kost., *Czekanowskia* ex gr. *rigida* Heer, *Pityophyllum* ex gr. *nordenskioldii* (Heer) Nath., and *Carpolithes heeri* Tur.-Ket.	0.15
28.	Sandstones, medium-grained, horizontally bedded	1.70
Thickness of the section opened part		13.21

The Upper Prisayanskaya Subformation outcrop within the Topka creek valley is presented by 18-meter-thick section having a rhythmic structure (**Figure 2**). The bases of rhythms are composed of quartz fieldspathic coarse- and medium-grained sandstones, siltstones, and mudstones with thin (0.1–0.3 m) coal bands. *Coniopteris murrayana* (Brongn.) Brongn., *C. maakiana* (Heer) Pryn. emend. Kiritch. et Trav., *Cladophlebis williamsonii* Brongn., *Cl. haiburnensis* (L. et H.) Sew., *Czekanowskia* ex gr. *rigida* Heer, *Phoenicopsis* ex gr. *angustifolia* Heer, *Phoenicopsis* ex gr. *speciosa* Heer, and *Pityophyllum* ex gr. *nordenskioldii* (Heer) Nath. are revealed among plant remains.

The Upper Prisayanskaya Subformation is exposed on the left bank of the Irkut river, 1.5 km below the Pionersk settlement (**Figure 2**). Outcrop of 800 m length has a rhythmic structure. The rhythm bases are composed of quartz fieldspathic coarse- and medium-grained sandstones (3–4 m) alternating upward with siltstones and mudstones with thin (0.1–0.2 m) coal bands. The following plant remains of *Coniopteris maakiana* (Heer) Pryn. emend. Kiritch. et Trav., *C.* cf. *spectabilis* Brick, *Coniopteris* sp., *Cladophlebis williamsonii* Brongn., *Cl. williamsonii* Brongn. var. *punctata* Brick, *Cl. nebbensis* (Brongn.) Nath., *Raphaelia diamensis* Sew., *Raphaelia tapkensis* (Heer) Pryn. emend. Kost., *Ginkgo* ex gr. *sibirica* Heer, *Sphenobaiera* ex gr. *czekanowskiana* (Heer) Flor., *Czekanowskia* ex gr. *rigida* Heer, *Phoenicopsis angustifolia* Heer, *Pityophyllum* ex gr. *nordenskioldii* (Heer) Nath., *Ixostrobus heeri* Pryn., *Carpolithes deplanatus* Pryn., and *Stenorhachis* sp. are identified in siltstones and fine-grained sandstones.

The Kudinskaya Formation is located in the northeast of the Irkutsk city within the Kuda trough, overlaps discordantly with the Prisayanskaya Formation. The formation is characterized by rubbly pebbled conglomerates and coarse-grained sandstones, thin interlayers of siltstones and mudstones are rare. Thickness of the formation is 50–80 m. Identifiable remains of plants from the Kudinskaya Formation have been long unknown. In 2014 we studied the stratotype of Kudinskaya Formation which is located at 2 km to the north of the Zherdovka village. Clastic rocks opened within these outcrops contain thin interlayers of siltstones and mudstones. The following fossil plants: *Equisetites* sp., *Coniopteris maakiana* (Heer) Pryn. emend. Kiritch. et Trav., *Czekanowskia curta* Kiritch. et Samyl., *Czekanowskia rigida* Heer, *Leptostrobus laxiflora* Heer, *Pityophyllum* ex gr. *nordenskioldii* (Heer) Nath. *Carpolithes cinctus* Nath., and *C. minor* Pryn. are found in mudstones [13, 26].

4. Discussion

The stratigraphic scale developed for Jurassic sediments of Western Siberia has been followed in the comparison of Jurassic sediments of the Irkutsk Basin with the adjacent regions of Western Siberia [3]. This scheme is comprehensively reasonable and can be considered as a comparative standard for large stratigraphic correlations and age dating of continental sections within Siberia.

Summarizing paleobotanical review of the aforementioned sections, it should be noted that plant assemblage identified in Lower Cheremkhovskaya Subformation is characterized by uncommon ferns *Cladophlebis haiburnensis* (L. et H.) Sew. and conifers *Podozamites eichwaldii* Pryn. var. *minor* Pryn., *Schizolepis* sp. Ginkgoales and Leptostrobales are presented by species *Sphenobaiera* ex gr. *czekanowskiana* (Heer) Flor. and *Czekanowskia* ex gr. *rigida* Heer, the latter one dominates among them. The taxonomic composition of the flora of the lower part of the Lower Cheremkhovskaya Subformation is comparable with Jagel'nyi fossil plant assemblage from the Lower Jurassic sediments of Western Siberia [21] (**Table 2**). The age of the Lower Cheremkhovskaya Subformation is determined within the framework of the second half of the Early Jurassic (Pliensbachian).

Various representatives of genus *Equisetites*: *E. lateralis* (Phill.) Phill. and *E. asiaticus* Pryn. are revealed in all studied sections of Middle and Upper Cheremkhovskaya Subformations. Genus *Cladophlebis* is presented by species *C. williamsonii* Brongn. and *Cl. haiburnensis* (L. et H.) Sew. that are abundant. *Czekanowskia rigida* Heer, *Cz. baikalica* Kiritch., *Sphenobaiera czekanowskiana* (Heer) Flor., *S. vigentis* Kiritch. et Bat., *Pityophyllum* ex gr. *nordenskioldii* (Heer) Nath. Реже встречаются *Lycopodites* sp., *Lobifolia nana* A. Frol., *Ginkgo* ex gr. *sibirica* Heer, *Pseudotorellia paradoxa* Dolud., *Phoenicopsis* ex gr. *angustifolia* Heer, *Leptostrobus laxiflorus* Heer, *Elatocladus manchuricus* (Yok.) Uabe, *Carpolithes cinctus* Nath., *C. minor* Pryn., *Carpolithes* sp., *Ixostrobus heeri* Pryn., *Ix. grandis* Tesl., and *Schizolepis follinii* Nath. play the role of cosmopolites among gymnosperms. *Hausmannia crenata* (Nath.) Maell., *Raphaelia diamensis* Sew., *Ginkgo concinna* (Heer) Sew., *Podozamites* cf. *lanceolatus* (L. et H.) Schimp., and *P.* cf. *eichwaldii* Pryn. var. *major* Pryn. are presented by several finds.

In whole, the flora composition of Middle and Upper Cheremkhovskaya Subformations is very consistent in outcrops that allow us to consider it as Cheremkhovo fossil plant assemblage

http://dx.doi.org/10.5772/intechopen.69572

[7]. *Equisetites lateralis* (Phill.) Phill., *E. asiaticus* Pryn., *Cladophlebis haiburnensis* (L. et H.) Sew., *Sphenobaiera czekanowskiana* (Heer) Flor., *S. vigentis* Kiritch. et Bat., *Czekanowskia baikalica* Kiritch. et Samyl., and *Cz. rigida* Heer whose remains dominate in the sections of Cheremkhovskaya Formation are typical of the assemblage (**Table 1**). Section of Middle

Fossil plant assemblage	Fossil plants \ Locations	Cheremchovskaya Formation						Prisayanskaya Formation							Kudinskaya Formation
		Middle Subformation			Upper Subformation										
		3	4	5	6	7	8	9	10	11	12	13	14	15	16
Prisayan	*Lycopodites baikalensis*											+			
	Lycopodites subulifolius										+				
	Coniopteris maakiana							□		□	□	□	□	□	□
	Coniopteris murrayana				+				□	□	□	□	□		
	Coniopteris spectabilis									+				+	
	Coniopteris cf. *sachsii*							+							
	Cladophlebis williamsonii	+	+				+	□	□	□	□	□	□	□	
	Cladophlebis nebbensis								□	+	□	□		□	
	Cladophlebis kanskiensis											+			
	Raphaelia diamensis	+								□	□	□		□	
	Raphaelia tapkensis										+	+		+	
	Baiera majaea									+					
	Ginkgo tapkensis									□					
	Phoenicopsis angustifolia							□	□	□	□	□	□	□	
	Phoenicopsis samylinae								□			□			
	Phoenicopsis dentata							+							
	Phoenicopsis irkutensis							□	□			□			
	Phoenicopsis markovitchiae							+							
Cheremchovo	*Lycopodites tenerrimus*				+										
	Lycopodites trichiatus				+										
	Equisetites lateralis	□	□												
	Equisetites asiaticus	□	□												
	Lobifolia nana			+						+		+	+		
	Cladophlebis haiburnensis	□	□				+			+		+		+	
	Sphenobaiera czekanowskiana	□	□		□	□	□								
	Sphenobaiera vigentis	□				□	□								
	Czekanowskia baikalika	□	□	□		□	□				+			+	
	Czekanowskia rigida	□	□	□	□	□	□								

Note: Numbering of fossil plan occurrences is given in **Figure 1**. □—Species dominates in burials, +—species is present in burials.

Table 1. Taxonomic composition and stratigraphic distribution of fossil plant assemblages within Jurassic sediments of the Irkutsk Basin.

General stratigraphic scale			Regional stratigraphic units															
			West Siberia (Kiritchkova et al, 2005)			Kuznetsk Basin (Kiritchkova et al, 2005)				Kansk Basin (Kiritchkova et al, 2005)				Irkutsk Basin (This work)			
Sustem	Series	Stage	Phytohorizon	FPA	Floral beds	Horizon	Formation	FPA	Floral beds	Horizon	Formation	FPA	Floral beds	Formation	Subformation	FPA	Floral beds
Jurassic	Middle	Bajocian	Tomskii	Azharminskii	Equisetites lateralis, Leptotoma batjaevae, Czekanowskia rigida, Phoenicopsis varia	Chernojetapskii	Tersyuk				Borodinskaya	Borodinskii	Raphaelia diamensis Czekanowskia teslenkoi, Cz. eugenia				
		Aalenian		Verhnepeshkovskii	Coniopteris maakiana, Ginkgo ananievii, Czekanowskia irkutensis, Phoenicopsis angustifolia, Ph. markovitchae, Kanskia		Osinovka	Etapskii	Equisetites lateralis, Coniopteris simplex, Cladophlebis suluctensis, Phoenicopsis cognata, Ph. markovitchae	Itatskii	Kamalinskaya	Rybinskii	Phoenicopsis markovitchiae, Ph. irkutensis, Czekanowskia kanensis, Ginkgo abaniensis, Kanskia	Kudinskaya Prisajanskaya	Upper Lower	Prisajan	Coniopteris maakiana, C. myrrayana, Raphaelia diamensis, R. tapkensis, Phoenicopsis angustifolia, Ph. cognata, Ph. irkutensis ?
	Lower	Toarcian	Urengojskii	Nizhnenovogodnii	Phlebopteris, Ginkgo sibirica, Sibiriella, Phoenicopsis irkutensis, Czekanowskia jenissejensis	Bungarapskii		Kamzasskii	Cladophlebis williamsonii, Sphenobaera vigentis, Czekanowskia baikalica	Perejaslovskii	Ilanskaya		?	Cheremhovskaya	Upper Middle	Cheremhovo	Equisetites lateralis, E. asiaticus Cladophlebis haiburnensis, Sphenobaiera vigentis, Czekanowskia baikalica, Cz. rigida
												Abanskii	Phoenicopsis cognata, Czekanowskia obiensis, Cz. rigida				
		Pliensbachian		Jagel'nyi	Neocalamites, Sphenobaiera magnifolia, Ginkgo tapkensis, Phoenicopsis cognata, Florinia (?)		Abasheva Raspadskaya	Abashevskii	Equisetites turgaicus, Pterophyllum tomiensis, Pituospermum maakianum, Samaropsis tersiensis, Etapia		Perejaslovskaya				Lower		Cladophlebis haiburnensis, Sphenobaiera ex gr. czekanowskiana, Ixostrobus grandis
					?				?								

Note: FPA—Fossil plant assemblage.

Table 2. Correlation scheme of Jurassic sediments of the Irkutsk Basin and adjacent areas on paleobotanical data.

Cheremkhovskaya Subformation of the Cheremkhovo hard coal deposit is a key for layers of the assemblage. Degree of development of the Cheremkhovo assemblage floras is comparable with floras of Novogodnii (Western Siberia) and Kamzasskii (Kuznetsk Basin) assemblages

(**Table 2**). Thus, the age of Middle and Upper Cheremkhovskaya Subformations is determined in the range of Toarcian.

The performed paleobotanical analysis of fossil plants found within Prisayanskaya and Kudinskaya Formations suggests that the time of their accumulation concurred with the flourishing of ferns of the genera *Conipteris* (*C. maakiana* (Heer) Pryn. emend. Kiritch. et Trav., *C. murrayana* (Brongn.) Brongn., *C. spectabilis* Brick, *C.* cf. *sachsii* Tesl.), *Cladophlebis* (*Cl. argutula* (Heer) Font., *Cl. haiburnensis* (L. et H.) Brongn., *Cl. kanskiensis* Kost., *Cl. nebbensis* (Brongn.) Nath., and *Cl. williamsonii* Brongn.) and *Raphaelia* (*R. diamensis* Sew. and *R. tapkensisi* (Heer) Pryn. emend Kost.). *Coniopteris maakiana* (Heer) Pryn. emend. Kiritch. et Trav., *Raphaelia diamensis* Sew., and *Cladophlebis nebbensis* (Brongn.) Nath. are dominant among them. *Coniopteris murrayana* (Brongn.) Brongn. и *Raphaelia tapkensis* (Heer) Pryn. emend Kost. occur rarely. Genus *Phoenicopsis*, presented by six species: *Ph. angustifolia* Heer, *Ph. samylinae* Kiritch. et Moskv., *Ph. irkutensis* Dolud. et Rasskaz., *Ph. cognata* Kiritch., *Ph. dentata* Pryn., and *Ph. markovitchiae* Kiritch. et Schischk. presented in majority of occurrences, is the most diverse among gymnosperms (**Table 1**). *Czekanowskia* genus is presented by species *Czekanowskia curta* Kiritch. et Samyl, *Cz. irkutensis* Kiritch. et Samyl, and *Cz. rigida* Heer. *Ginkgo* ex gr. *sibirica* Heer. and *G. tapkensis* Dolud. et Rasskaz. are frequent. Besides them, *Hepaticites arcuatus* (L. et H.) Harris, *Lycopodites baikalensis* A. Frol., *L. subulifolius* A. Frol. et Mash., *Phyllotheca sibirica* Heer, *Hausmannia crenata* (Nath.) Mael., *Lobifolia lobifolia* (Phill.) Rasskaz. et E. Leb., *Anomozamites lindleyanus* Schimp., *Sphenobaiera* ex gr. *czekanowskiana* (Heer) Florin, *Pseudotorellia* cf. *ensiformis* (Heer) Dolud., *P.* cf. *paradoxa* Dolud., *Taxocladus ketovae* Tesl., *Elatocladus manchuricus* (Yok.) Yabe, *Pityophyllum* ex gr. *nordenskioldii* (Heer) Nath., *Ixostrobus heeri* Pryn., *Carpolithes cinctus* Nath., *C. deplanatus* Pryn., *C. heeri* Tur.-Ket., *C. minor* Pryn., *Schizolepis follinii* Nath., *Samaropsis rotundata* Heer, and *Stenorachis* (?) *clavata* Nath. are found within the Prisayanskaya Formation.

The flora composition of Prisayanskaya and Kudinskaya Formations shows good horizontal consistency, and we consider it as Prisayan fossil plant assemblage. Species *Coniopteris maakiana* (Heer) Pryn. emend. Kiritch. et Trav., *C. murrayana* (Brongn.) Brongn., *C. spectabilis* Brick., *Cladophlebis nebbensis* (Brongn.) Nath., *Raphaelia diamensis* Sew., *R. tapkensis* (Heer) Pryn. emend Kost., *Phoenicopsis angustifolia* Heer, *Ph. cognata* Kiritch., and *Ph. irkutensis* Dolud. et Rasskaz., the remains of which prevail in the sections of Prisayanskaya and Kudinskaya Formations (**Table 1**), are characteristic of this assemblage [7, 13]. Composition of Prisayan assemblage allows it to compare to Verkhnepeshkovskii (Western Siberia), Rybinskii (Kansk Basin) and Etapskii (Kuznetsk Basin) fossil plant assemblages and date including sediments to the beginning of Middle Jurassic (Aalenian) (**Table 2**). Outcrops situated near the Smolenshchina settlement and on the left bank of the Irkutsk water reservoir serve as key sections for Prisayan assemblage.

Plant remains from the sections of Upper Prisayanskaya Subformation opened within the Topka river were previously compared, according to the level of flora development, with Azharminskii fossil plant assemblage traced in Tyumen'skaya Formation of Ob'-Tazovskaya area of Western Siberia [21]. Azharminskii assemblage is characterized by renewal of species composition of genera *Coniopteris*, *Czekanowskia*, and *Phoenicopsis*. Genus *Coniopteris* is replenished with species *C. burejensis* (Zaless.) Sew., *C. (Birissia?) depensis*

E. Leb. Species *C. hymenophylloides* (Brongn.) Sew., *C. simplex* (L. et H.) Harris, *C. vsevolodii* E. Leb. Occur everywhere. Genus *Phoenicopsis,* presented by 10 species, is the most diverse among Leptostobales. Species *Ph. taschkessiensis* Krasser and *Ph. mogutchevae* Kiritch. et Trav. are the first to occur among them. The most abundant are species *Ph. samylinae* Kiritch. et Moskvin., *Ph. sibirica* Kiritch. et Trav., and *Ph. varia* Kiritch. et Trav. The age of sediments including Azharminskii fossil plant assemblage is determined in the range of Bajocian stage. Analysis of flora taxonomic composition from the sections of the Topka river revealed in it the lack of all new species of genera *Coniopteris* and *Phoenicopsis* typical of Azharminskii assemblage (**Table 1**). Thus, according to the available data, there is no Bajocian flora within the Irkusk Basin.

5. Conclusion

The results of lithologic and paleobotanical investigations of Lower and Middle Jurassic sediments of the Irkutsk Basin allowed to draw the following conclusions.

Two uneven-aged fossil plant assemblages: Cheremkhovo for the Middle and Upper Cheremkhovskaya Subformations and Prisayan for the Prisayanskaya and Kudinskaya Formations were identified.

Species *Equisetites lateralis* (Phill.) Phill., *E. asiaticus* Pryn., *Cladophlebis haiburnensis* (L. et H.) Sew., *Sphenobaiera czekanowskiana* (Heer) Flor., *S. vigentis* Kiritch. et Bat., *Czekanowskia baikalica* Kiritch. et Samyl., and *Cz. rigida* Heer were dominated in sections of Middle and Upper Cheremkhovskaya Subformations. Level of development of Cheremkhovo assemblage is comparable with flora of Nizhnenovogodnii (Western Siberia) and Kamzasskii (Kuznetsk Basin) assemblages. Therefore, the age of Middle and Upper Cheremkhovskaya Subformations including Cheremkhovo assemblage is determined by the end of Early Jurassic (conventionally, Toarcian).

Species *Coniopteris maakiana* (Heer) Pryn. emend. Kiritch. et Trav., *C. murrayana* (Brongn.) Brongn., *C. spectabilis* Brick., *Cladophlebis nebbensis* (Brongn.) Nath., *Raphaelia diamensis* Sew., *R. tapkensis* (Heer) Pryn. emend Kost., *Phoenicopsis angustifolia* Heer, *Ph. cognata* Kiritch., and *Ph. irkutensis* Dolud. et Rasskaz. are typical of Prisayan assemblage. The identified species composition of the Prisayan assemblage allowed to compare it with Verkhnepeshkovskii (Western Siberia), Rybinskii (Kansk Basin), and Etapskii (Kuznetsk Basin) fossil plant assemblages and to establish the formation age at the beginning of the Middle Jurassic (Aalenian).

Acknowledgements

The study was financially supported by the Russian Foundation for Basic Research (16-35-60005).

Author details

Andrey Olegovich Frolov, Nikolay Ivanovich Akulov* and Irina Mikhailovna Mashchuk

*Address all correspondence to: akulov@crust.irk.ru

Institute of the Earth's Crust Siberian branch of Russian academy of science, Irkutsk, Russia

References

[1] Akulov NI, Frolov AO, Mashchuk IM, Akulova VV. Jurassic deposits of the southern part of the Irkutsk sedimentary basin. Stratigraphy and Geological Correlation. 2015;**23**(4):387-409. DOI: 10.1134/S0869593815040036

[2] Decisions of the III Regional Interagency Stratigraphic Meeting on Mesozoic and Cenozoic of Central Siberia: MSK USSR. Novosibirsk: Siberian Scientific-research institute of geology, geophysics and mineral raw material (SNIIGGiMS); 1981. p. 91 [in Russian]

[3] Decisions of the VI Interdepartmental Stratigraphic conference on Consideration and Adoption of Emended Stratigraphic Schemes of Mesozoic Deposits of Western Siberia, Novosibirsk, 2003. Novosibirsk: Siberian Scientific-research institute of geology, geophysics and mineral raw material (SNIIGGiMS); 2004. p. 114 [in Russian]

[4] Doludenko MP, Rasskazova YS. Ginkgoales and Czekanowskiales of the Irkutsk basin. In: Vachrameev VA, editor. Mesozoic Plants of Eastern Siberia. Moscow: Nauka; 1972. pp. 7-43 [in Russian]

[5] Ilyina VI. Jurassic Palynology of Siberia. Moskow: Nauka; 1985. p. 237 [in Russian]

[6] Frolov AO. Schizolepis mashchukae sp. nov.—A new species from Middle Jurassic deposits of the Irkutsk coal basin (Eastern Siberia). Herald of the Tomsk State University. 2012;**362**:194-196 [in Russian]

[7] Frolov AO. Early and Middle Jurassic plant communities of the Irkutsk coal basin [thesis]. Tomsk: Publishing House "Pozitiv-NB"; 2013 [in Russian]

[8] Frolov AO, Mashchuk IM. Field atlas of the Jurassic flora of the Irkutsk coal basin. Irkutsk: Institute of the Earth's Crust SB RAS; 2014. p. 108 [in Russian]

[9] Frolov A, Mashchuk I. A new species of extinct genus Lycopodites from Lower to Middle Jurassic sediments of Irkutsk coal basin (Eastern Siberia). Global Geology. 2014;**1**:1-10

[10] Frolov AO, Mashchuk IM. A new fern from Lower Jurassic sediments of the Irkutsk coal basin (Eastern Siberia). Paleontological Journal. 2015;**49**(4):424-428. DOI: 10.1134/S0031030115040073

[11] Frolov AO, Mashchuk IM. The first record of the species Cladophlebis kanskiensis Kost. (fern) in Middle Jurassic sediments of the Irkutsk basin (Eastern Siberia, Russia). The Bulletin of Irkutsk State University. Series "Earth Sciences". 2016;**16**:128-136 [in Russian]

[12] Frolov AO, Mashchuk IM. Rare conifers from the Jurassic sediments of the Irkutsk coal basin (Eastern Siberia, Russia). The Bulletin of Irkutsk State University. Series "Biology, Ecology". 2016;**15**:25-36 [in Russian]

[13] Frolov AO, Mashchuk IM, Arzhannikova AV. First paleobotanic findings from Kudinskaya and Tal'tsinskaya formations (Irkutsk coal basin) and their stratigraphic importance. In: Structure of Lithosphere and Geodynamics: Proceedings of XXVI All-Russian Youth Conference; 20-25 April 2015; Irkutsk. Irkutsk: Institute of the Earth's Crust SB RAS; 2015. pp. 204-205 [in Russian]

[14] Heer O. Beiträge zur Jura-Flora Ostsibiriens und des Amurlandes. Flora Fossils Arctica. 1876;**4**:1-122

[15] Heer O. Jurassic flora of the Irkutsk guberniya and the Amur Krai. Proceedings of the Siberian Expedition of the Russian Geographical Society. Physical Department, Vol. III, Iss. 2. St. Petersbourg; 1878. p. 134 [in Russian]

[16] Heer O. Beiträge zur fossilen Flora Ostsibiriens und des Amurlandes. Flora Fossils Arctica. 1878;**5**:58

[17] Kiritchkova AI, Batyaeva SK, Bystritskaya LI. Phytostratigraphy of Jurassic deposits from the south of Western Siberia. Moscow: Nedra; 1992. p. 216 [in Russian]

[18] Kiritchkova AI, Travina TA. On sphenopteroidal ferns from the Jurassic of the Irkutsk basin (Ust'-Baley and Kaja localities). Paleontological Journal. 1993;**4**:106-114 [in Russian]

[19] Kiritchkova AI, Kostina EI, Travina TA. New species of Osmunda L. from the Jurassic deposits of the Irkutsk coal basin. Paleontological Journal. 1999;**2**:83-89 [in Russian]

[20] Kiritchkova AI, Travina TA. Phytostratigraphy of Jurassic coal-bearing deposits of the Irkutsk basin. Stratigraphy and Geological Correlation. 2000;**8**(6):89-102 [in Russian]

[21] Kiritchkova AI, Kostina EI, Bystritskaya LI. Phytostratigraphy and flora of Jurassic deposits of the Western Siberia. St. Petersburg: Nedra; 2005. p. 378 [in Russian]

[22] Khakhlov VA. Fossil plants of the Irkutsk coal basin. Proceedings of the Siberian Branch of the Geological Committee Tomsk. 1924;**4**:29 [in Russian]

[23] Korovin MK. Cheremkhovo hard coal basin. Proceedings of the Siberian Branch of the Geological Committee Tomsk. 1922;**2**(4):64 [in Russian]

[24] Kostina EI. Jurassic flora of the Kansk coal basin. Moscow: GEOS; 2004. p. 166 [in Russian]

[25] Krassilov VA, Bugdaeva EV. Gnetalean plants from the Jurassic of Ust-Balej, East Siberia. Review of Palaeobotany and Palynology. 1988;**53**:359-374

[26] Mikheeva EA, Demoterova EI, Frolov AO, Arzhannikova AV, Arzhannikov SG, Cherkashina TY, Ivanov AV. Provenance change in Irkutsk coal-bearing basin by

paleontology, geochemic and Sm-Nd isotopic data. Stratigraphy and Geological Correlation. 2017;**25**:3. DOI: 10.7868/S0869592X1703005X [in press]

[27] Prinada VD. Mesozoic flora of Eastern Siberia and Transbaikalia. Moscow: Gosgeoltekhizdat; 1962. p. 368 [in Russian]

[28] Skoblo VM, Lyamina NA, Rudnev AF, Luzina IV. Continental Upper Mesozoic of Cisbaikalia and Transbaikalia (stratigraphy, sedimentation conditions, correlation). Novosibirsk: Publishing House of the SB RAS; 2001. p. 332 [in Russian]

[29] Yermolaev DI. On the problem about age of coal-bearing deposits of the Irkutsk coal basin. Materials on geology and minerals of Eastern Siberia. Publishing of the Irkutsk Geological Man. 1958;**3**:17-21 [in Russian]

Chapter 6

Stratigraphic Unconformities: Review of the Concept and Examples from the Middle-Upper Paleozoic

Pavel Kabanov

Additional information is available at the end of the chapter

http://dx.doi.org/10.5772/intechopen.70373

Abstract

Only about 10% of geologic time is imprinted in sedimentary strata and the rest is hidden in non-depositional or erosional surfaces called unconformities. Stratigraphic unconformities (disconformities) are principal bounding surfaces in sequence stratigraphy, which a geologist would easily identify in the outcrop but frequently overlook in the subsurface unless core is available. The proportion of disconformities that are misidentified or overlooked in subsurface stratigraphy is quite large, which puts a warning sign on simplistic sequence stratigraphic models. The amount of time imprinted in disconformities can be evaluated using relative weathering maturity of the subaerial profile, cyclostratigraphic calibration, absolute dating, and biostratigraphy. However, using biostratigraphy alone is never enough as biostratigraphic gaps tend to fill with increasing data coverage. Identification of paleo-vadose zones and subaerial exposure profiles is regarded as critical for finding stratigraphic unconformities and is the only approach in strata where geophysically mappable fluvial systems are absent. Drowning unconformities are carbonate platform drowning surfaces that usually produce distinct reflection horizons and have better stratigraphic value in the subsurface than platform-embedded subaerial unconformities. This discussion is supported by examples of subaerial disconformities from the Devonian, Carboniferous, and Permian of Canada and Russia and with an example of a geographically extensive mid-Devonian drowning unconformity from Northwestern Canada.

Keywords: disconformities, paleosols, paleokarsts, vadose alteration, erosion, drowning unconformities, sequence stratigraphy, subsurface identification

INTECH
open science | open minds

1. Introduction

1.1. Definition of the subject

It is generally accepted that only about 10% of the geologic time is recorded in the sedimentary rocks, whereas 90% is collapsed into non-deposition, alteration, and erosion surfaces collectively called unconformities [1]. Of these diverse surfaces with time value ranging from minutes to hundred millions of years, only those of practical use, that is, traceable on a scale exceeding one outcrop and marked with distinct diagnostic features are discussed below.

1.2. Growth of the concept

This subheading, borrowed from Dunbar and Rodgers [2], brackets 230 years of unconformity research counting from recognition of an angular unconformity in late 1780s [3]. The word *unconformity* was adapted from German geology 3 decades later [4, p. 48] and until the mid-nineteenth century pertained to angular stratal discordances. Awareness of geologic time gaps between parallel bed sets, normally accompanied by signatures of erosion, emerged in late nineteenth century (e.g., [5]) under the influence of Charles Darwin's conclusion on the principal incompleteness of the stratigraphic record [6]. As most recently reviewed by Miall [1, 7], such stratigraphic breaks between parallel strata were classified into "unconformities Type a" by Blackwelder [8] and shortly after that named *disconformities* [9]. The other two types of unconformities of Blackwelder [8] were (b) contact between rocks of wholly unlike origin (for example, sandstone resting upon granite); and (c) angular discordance of beds with or without difference in lithologic character. Type (c) is the classical *angular unconformity* of James Hutton, and type (b) was named *nonconformity*. The latter term was coined by Pirsson and Schuchert [10] and refined into modern usage by Dunbar and Rodgers [2]. Surfaces between parallel bed sets recording time gaps but not bearing signs of erosion were named *paraconformities*, as opposed to erosion-marked *disconformities* [2]. However, the difference between the disconformity and paraconformity more often appears in the ability to recognize erosion and evolves with tools and methods. Here, the term stratigraphic unconformity is used as an equivalent of disconformity. Barrell [11] also coined a term *diastem* that became adapted for the time value of a sedimentation gap at an unconformity. Being most easily identified features, angular unconformities and nonconformities are excluded from further discussion.

Disconformity-bounded packages of sedimentary rocks, called cycles, cyclites, cyclothems, allostratigraphic units, and most commonly sequences, remained in focus for many decades, generating an impressive development of concepts, terminology, and discussion on local vs. global controls of base or sea level fluctuations [1, 2, 12–24]. It should be noted that sequence stratigraphy significantly expanded definition of sequences by including both disconformities and their correlative surfaces (conformities) in more complete basin-centered sections [14]. Sequence stratigraphy is reviewed in this book but is not the focus of this contribution.

http://dx.doi.org/10.5772/intechopen.70373

2. Disconformities at seismic resolution

Disconformities mostly show concordant stratal relationships below and above the surface. They are identifiable on seismic sections if the subaerial exposure allowed for development of significant relief and/or seismic-scale incised fluvial channels [20, 25]. Incised valleys form during base level fall and become filled when base level rises [26]. Fluvial incisions are deeper and favorable for seismic mapping where they cut into an uplifted plain (**Figure 1A**), plateau, or across a shelf break (**Figure 1B**), but may not be identifiable in paleo-hinterlands with a shallow base of erosion. Transgressive tide and wave abrasion of coasts, estuaries, and shoreface are able to modify the configuration of subaerial surfaces and any terrestrial sediment accumulated on it. The surfaces produced by such an abrasion are called ravinement surfaces [17, 22, 23]. The depth of transgressive erosion greatly varies depending on induration of the exposed sediment, on the wave and tide energy of a transgressing sea, and on the slope angle of the eroded sediments. While oceanic abrasion may cut down to tens of meters into seashore cliffs, plain lands characteristic of epicontinental sedimentary environments may show negligible transgressive stripping and delicate topsoil parts of weathering profiles largely preserved (e.g., [27]).

Seismic and hands-on-rock unconformities are not the same, and the proportion of false seismic unconformities is greater than was thought by Vail et al. [15]. Situations where time lines converge into a condensed section but portrayed as an onlap-offlap surface, or pseudo-unconformities envisioned from a surface of major lithological contrast, are very common misinterpretations [30, 31]. Difficulty in recognition of subaerial unconformities in the subsurface led to proposal of an alternative *genetic stratigraphic sequences* bounded by "maximum flooding surfaces" or condensed sections [16]—however, the concept of very limited use today.

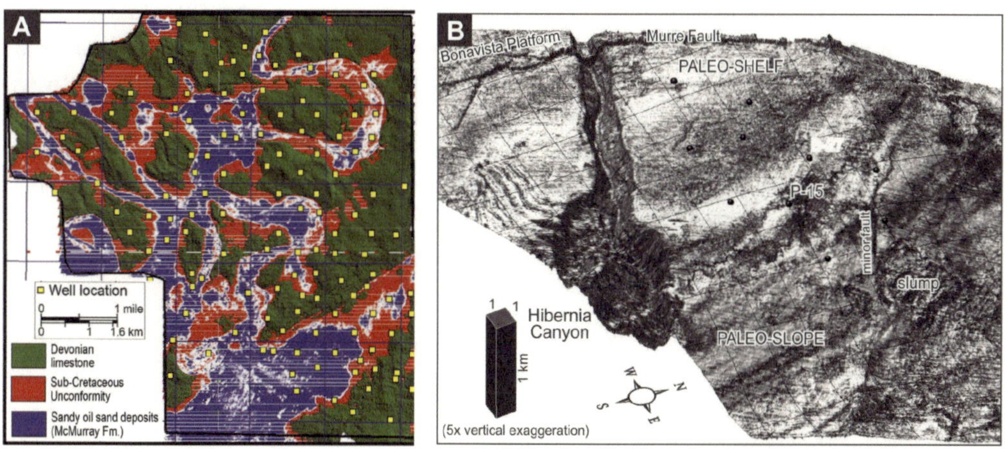

Figure 1. Examples of mature unconformity surfaces with fluvial incisions visualized in 3D seismic models: (A) A 225 m subsurface slice (above sea level) along sub-Cretaceous unconformity visualizing high-sinuosity channels cut into weathered Devonian limestone, eastern Athabasca oil sands, Alberta (formally modified from [28]). (B) Hibernia Canyon cut through shelf edge during latest Campanian-earliest Danian (Cretaceous-Paleocene), Jeanne d'Arc Basin offshore Newfoundland, formally modified from Deptuck et al. [29].

Disconformities are also invisible with conventional seismic surveys in carbonate successions if (1) no reef crest or carbonate mound relief was produced in slowly subsiding platform setting; (2) transgressive carbonates deposited upon regressive carbonates with no impedance contrast produced; (3) sequences are thinner than reading resolution at a given pulse frequency; (4) fluvial channels or sufficient erosional relief did not develop and drainage of meteoric waters was entirely underground; and (5) stacked karst systems from successive paleo-aquifers overprint with no chance to trace particular karst horizons. The best present-day example of a carbonate plain where the day surface has a chance to be buried in such a hidden way is the Nullarbor Plain of Southwestern Australia [32]. This vast (~240,000 km^2) plain was exposed for the last 14 My since mid-Miocene time, yet remains exceptionally flat and riverless with extensive underground cave systems produced during several Tertiary lowstands, including the ongoing uplift [33].

3. Evaluation of hiatuses

Most terrains show a relief or a slope gradient where prolonged flooding is recorded in onlap patterns, as opposed to geologically momentary (rapid) inundation of plain lowlands. Hiatuses therefore tend to wedge toward basin centers on chronostratigraphic charts [15], as demonstrated by case studies where unconformities receive cross-basinal biostratigraphic control [29]. Hiatuses reveal more complex histories in settings of differential subsidence in areas of large-scale salt diapirism or in tectonically active regions (e.g., foreland flexural bulging and tilting). The eustatic vs. tectonic control over transgressions and regressions is a subject of long-lasting debate [13, 20]. Tectonic control of an unconformity between two parallel bed sets can be interpreted where unconformities show poor or no correlation to major lowstands of "global sea level curves" or where the hiatus is diachronous with bedrock and caprock younging in the same direction. For example, a major intra-Cretaceous disconformity of central-southern Italy is generally younging eastward from Late Albian to Late Turonian—earliest Coniacian as revealed with refined biostratigraphic control [34]. This unconformity hosts karst-associated economic bauxites and is locally composite with two bauxitiferous paleokarsts divided by Cenomanian limestone of various thickness and time value. This diachroneity was interpreted as the translation of the lithospheric bulge in response to compression from the distal orogeny along the Adria Plate margin [34].

Biostratigraphy is the oldest yet still master method of recognizing hiatuses by missing zones, which can be processed with a graphic correlation technique [7]. Other absolute dating methods like U-Pb ID-TIMS and cyclostratigraphy are reviewed in [7, 35]. Resolution of biostratigraphy varies with the group employed, paleogeographic position, and the geologic age. The latter controls biostratigraphic resolution to a significant extent by cosmopolitism vs. provincialism of marine faunas. High cosmopolitism is characteristic of greenhouse periods with circum-tropical seaway connections, whereas provincialism is favored by forcing of the Earth into icehouse mode and shutdowns of low-latitude seaway connections, as likely happened during Pennsylvanian-Permian assembly of Pangaea [36, 37].

Noteworthy here are historically recognized but apparently non-existent disconformities. Usually, these "legacy hiatuses" heavily rely on biostratigraphy. An example is given by the "Late Middle Devonian unconformity" of the Mackenzie Corridor of Northwestern Canada. This unconformity was interpreted by Hume and Link [38] from the sharp thickness fluctuations and restricted spatial distribution of the Hare Indian and Ramparts formations, which was seen as a result of erosion prior to deposition of the black siliceous shale of the Canol Formation (**Figure 2**; [46]). A debate on the validity of this hiatus lasted ever since. The hiatus has been supported by the assignment of the upper Hare Indian Formation to the undifferentiated *varcus* conodont zone (=*rhenanus-ansatus* in **Figure 2**), whereas the lower part of Canol Formation was dated by conodonts as the Lower *asymmetrica* (≈*transitans-falsiovalis* on **Figure 2**) with speculative extension of the Canol base into the lowermost *asymmetrica* or present-day *norrisi* zone [39, 40]. *Hermanni-disparilis* interval was allegedly missing (**Figure 2**). However, scarce conodont data from the Ramparts limestone suggested its age range from the upper Hare Indian equivalent

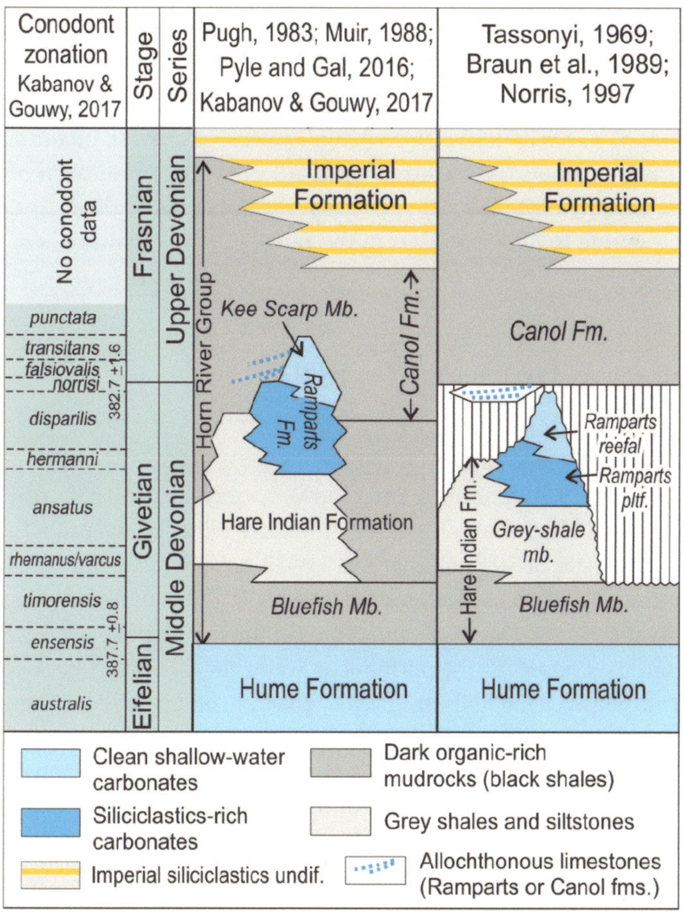

Figure 2. Legacy "Late Middle Devonian unconformity" on a simplified table of formations of Mackenzie Valley and Peel area; the unconformity advocated (right column) vs. discarded (left column).

to the *asymmetrica* zone [41]. Other workers argued that the Canol base is a conformity [42, 43] and indicated Ramparts-Canol interfingering in allochthonous debris units [41]. Nevertheless, this hiatus survived in the territorial table of formations until recently [44]. Decisive in retiring this hiatus are (1) updates in conodont data showing Canol base time gliding from the Frasnian *transitans-punctata* on top of Kee Scarp carbonate banks to the upper Givetian *norrisi* in off-bank depressions [44]; (2) carbonate-bank slope depositional setting of allochthonous bioclastic debris interfingering with laminated black shales; and (3) absence of any evidence of subaerial exposure or vadose processes, like oxidation of pyrites and organic matter and characteristic redistribution of Fe and Mn, prior to the onset of Canol deposition [45].

An absolute majority of disconformities are subzonal or do not bear index fossils immediately below or above. Relative proxy for the duration of a hiatus is the maturity of a paleosol profile, e.g., progression from entisols to any of mature soil profiles defined by soil taxonomy [47], or stages of calcrete development [48], but the ability to deconvolute time is quite limited: paleosol appearance is a multivariate product of exposure duration, precipitation regime, temperature, relief, availability and type of vascular vegetation (and other soil biota), and transgressive truncation, with variable masking of paleonvironmental signals by burial diagenetic overprints. Most tools of radiogenic dating used to reveal soil age are not applicable to deep-time examples because of short isotope decay lifetime. U-Pb dating of soil carbonates, based on U adsorbed in calcite lattice, was demonstrated to provide quantitative estimate of pedogenic processes as old as Carboniferous [49, 50]. Also, the production of He isotopes by α-decay of U, Th, and their intermediate decay species was used to develop a (U-Th)/He geochronometer that is able to date materials in the range of a few thousands of years to 4.5 Ga (see review in [51]).

4. Paleopedology and paleoaquifer studies

Soils blanket most of the terrestrial landscape [47]. As approximated by Landsat-based NARWidth model of North American river surface, only about 0.55% of the continent is covered with rivers [52]. This model includes measurements of rivers that are ≥30 m wide and extrapolated estimate of streams that are 1.6–30 m wide. Natural lakes and reservoirs excluding human-made impoundments cover about 3% of the World's land surface [53]. Therefore, at this momentary snap of the geologic time, it appears that no less than 96% of the land terrain is exposed above permanent water level if timed back to pre-industrial landscape. This emphasizes the high probability that a geologist will encounter in particular section a subaerial unconformity with a paleosol or what is left after its transgressive erosion, rather than non-pedogenized fluvial or lake sediments (**Figure 3**). This considers paramount importance to recognition of paleosols, regoliths, and former meteoric aquifers, although these are usually subtle features masked by diagenetic alteration and not readily picked with geophysical surveys.

Recognition of fossil soils in pre-Quaternary strata commenced early in nineteenth century with description of "dirt beds" with fossil wood stumps in the Upper Jurassic of the Dorset Coast

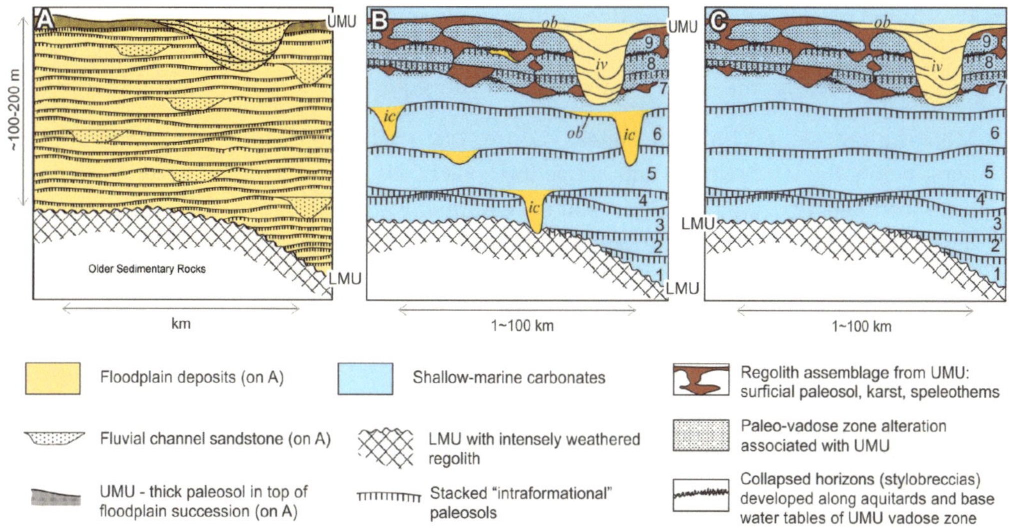

Figure 3. Conceptual expression of disconformities: (A) in floodplain succession, formally modified from Kraus [54]; (B and C) in shallow-marine carbonate succession; (B) deposited under wet climate with precipitation sufficient for open stream drainage; (C) deposited in drier climate with underground karst drainage. Paleo-vadose zones associated with "intraformational" unconformities (not shown here for simplicity) extend to various depth beneath paleosols, depending on base level fall, and frequently overprint; lower major unconformity (LMU) and upper major unconformity (UMU); incised fluvial channels (ic); a deep incised valley (iv) developed from UMU; and overbank (ob) floodplain deposits. Units in (B) and (C) bounded by disconformity surfaces are numbered; note that depending on researcher's knowledge, they may be described as sequences if subaerial unconformities are adequately characterized, as parasequences if surfaces known but their genesis is uncertain, or merged in one sequence bounded by LMU and UMU if surfaces remain below resolution with available tools.

[55–57]. Paleopedology (study of fossil soils) is a discipline nursed on soil science, specifically Quaternary soil chronosequences [58] and later expanded back in the geologic time, even to other planets, with adsorption of geologic and biotic evolutionary concepts [47, 57]. A short glossary of basic concepts is given below.

- **Paleosol** is an ancient soil or part of it that has been imprinted in the stratigraphic record [47, 51]. Soil is defined as (1) the medium rooted and modified by vascular plants (narrow pedologic definition; e.g., [59]) and (2) biologically and chemically active "excited skin" of the subaerial part of the Earth's crust (broad pedologic definition; [60]).

- **Regolith**, first defined by Merrill [61], refers to all the continental lithospheric materials above fresh (non-weathered) bedrock and including block of fresh rocks where they are interbedded with or enclosed by unconsolidated or weathered rock [62]. Regolith mantles the fresh rock and consists, from base to top, of saprock (patchily weathered rock), saprolith (pervasively weathered rock), and the soil (or paleosol in buried examples) where pedologic horizons are recognized [62]. The word *regolith* is also employed to describe weathered rock mantles of other planets and pre-land biota Earth [47]. The concept of *regolith* is sometimes considered vague, and in situ regoliths are taken as equivalent to soil *s.l.* [57].

- **Weathered crust** is a retiring wordage still in use in some geological schools of the World; it is applied to different scale features from thick lateritic mantles with bauxites to alteration rinds on pebbles; mantle-scale weathered crusts are equivalent to *in situ* regoloths *sensu* [62].
- **Meteoric vadose and phreatic zones.** Vadose zone applies to the diagenetic environment lying below the land surface and above the zone of saturation or water table where pore space contains both water and air (soil gas); rainfall waters percolate downward developing vadose patterns of dissolution, reprecipitation and alteration of host sediment. Meteoric phreatic zone applies to aquifer of permanent saturation below the water table and is divided from the vadose zone by a capillary fringe [63]. Concept of vadose and phreatic zones is employed to describe carbonate aquifers, subaerial alterations, and karst systems.
- **Aquifer** is a rock formation saturated with groundwater that is porous and permeable enough for sufficient debit to wells and springs; related are *aquicludes* and *aquitards*. Aquiclude is saturated but do not transmit groundwater; *aquitard* is a low-permeability or impermeable rock formation, usually strata, that confine water flow.
- **Critical zone** is a young concept referring to near-surface environment in which complex interactions involving rock, soil, water, air, and biota regulate the natural habitat and determine availability of life sustaining resources [64]; near-surface terrestrial environment in which resource availability is determined by interactions between the biosphere, geosphere, and atmosphere [51]; includes regolith, within- and below-regolith aquifers, fluvial systems, soils, and vegetation up to tree canopy [65].

Under different climate and hydrologic regimes, soil-forming processes create diverse soil profiles described by national and international soil classifications. The North American soil taxonomy [66] is the one that has earned greatest recognition in paleopedologic studies [47, 51]. Diagnostic criteria for pre-Quaternary paleosols and instrumental proxies for landscape and climate reconstructions are reviewed in [47, 51, 54, 67, 68].

Diagnostics of subaerial exposure profiles and paleokarst systems in carbonate rocks (**Figure 3B, C**) were developed by sedimentary geologists as a parallel story to paleopedology, which was driven by economic importance of karst as (1) hydrocarbon reservoir-making factor and (2) a host for bauxite [34] and rare metal accumulation [69]. This move has generated very practical terminology focused on horizons of high preservation potential, first of all caliches (calcretes) and paleokarsts, with various degree of reconciliation with the soil science lexicon [48, 69-71].

5. Drowning unconformities

Drowning unconformities are "maximum flooding surfaces" (= drowning surfaces *sensu* Posamentier and Allen [17]) specific for carbonate platforms.

In the subsurface, drowning unconformities usually make good seismic reflectors with basinal strata onlapping carbonate slopes and platform tops [72]. On the outcrop or core face, these contacts are characterized by condensed sections (e.g., shell concentrates) and non-deposition surfaces with hardgrounds sometimes impregnated with phosphate and/or glauconite [73, 74]. The surface of a drowned carbonate platform can be table-flat, rimmed by a reef crest (empty-bucket configuration), or outgrown with backstepped carbonate mounds or "pinnacles" [31, 75]. Drowning unconformities are within-trend drowning ("flooding") surfaces in sequence stratigraphy [20]. Fundamental genetic difference from subaerial unconformities renders certain reluctance in accepting them as formal sequence boundaries (e.g., [76]). However, drowning surfaces are more practical in subsurface surveys as they produce vivid onlap pattern on carbonate slopes and sharp impedance contrasts, in difference to feeble expression of subaerial disconformities embedded in carbonate platforms. Also, submarine corrosion can mimic subaerial karst to some degree, complicating its workflow recognition and leading to misinterpretations. These considerations led to proposal to legalize drowning unconformities in sequence stratigraphy as *Type 3 sequence boundaries* [18, 31].

Factors leading to demise and drowning of carbonate platforms are rapid relative sea level rise and/or carbonate production shutoff by eutrophic turbid waters, either loaded with siliciclastics or upwelled from deep ocean [72, 74], but these factors can only smother photozoan or tropical carbonate factories (*T-factories*; [31, 77]). Non-actualistic mud-mound carbonate factory can likely produce thick carbonate buildups in dimmed or aphotic settings and at elevated nutrient levels (*M-factories*; [31, 77]). Drowning unconformities are usually produced in settings of tectonic subsidence, e.g., in extensional rifted basins [78] or foreland basins [79–81]. It remains unclear to what extent high-amplitude eustatic rise of sea level, without aid of other factors, is capable of shutting off carbonate platforms. Another factor is the slowdown in ocean circulation under greenhouse condition of the Earth or even shutdown of thermal ocean circulation under extreme hothouse condition [82], which should lead to lateral expansion and shallowing of the oxygen minimum zone in the ocean (OMZ; [83]). The OMZ under such conditions should develop a thick and permanent euxinic environment in its core and should be able to rapidly shut down benthic carbonate factory across broad expanse of an ocean-facing carbonate shelf even in tectonically quiet setting. Such OMZ expansions are seen as the condition imprinted in severe form in oceanic anoxic events [84]. Possible link between synchronous and widespread demise of carbonate platforms and oceanic anoxic events has been indicated based on the Cretaceous "Selli event" (OAE 1a; [85]). Even during icehouse epochs, anoxic waters similar to those in present-day eastern tropical pacific OMZs were likely able to encroach far into interiors of epeiric seas during interglacial highstands and switch carbonate deposition to black phosphatic shales [86].

6. Case studies

6.1. Permo-Pennsylvanian of Sverdrup Basin, Canadian Arctic Archipelago

Eight subaerial unconformities define major sequences in the Pennsylvanian and Permian strata of the Sverdrup Basin [87–89]. These unconformities are considered to be subaerial surfaces of long duration (>1 My) bounding thick (100–1000 m) third-order sequences [90]. They are correlated across the basin in outcrops of the basin-margin facies belt. Five of these unconformities and their correlative surfaces were traced in the subsurface of Prince Patrick Island [90]. Recent re-examination of cores has confirmed the presence of subaerial exposure surfaces [91]. Subaerial profiles in cores are mostly decapitated by erosion (transgressive ravinements) but preserve features such as calcretic and ferric replacive deposits, *Microcodium*, root traces, and high-chroma ("red") mottling, which provide a clue for their interpretation (**Figure 4**). Of 49 short (<18.5 m) cores totaling 388 m of recovery, signatures of subaerial alteration were encountered in 8 (**Table 1**). Four of these cores intersect disconformity surfaces and one core penetrated the sub-Pennsylvanian angular unconformity into the Ellesmerian basement (Depot Island C-44, 2458.2 m MD).

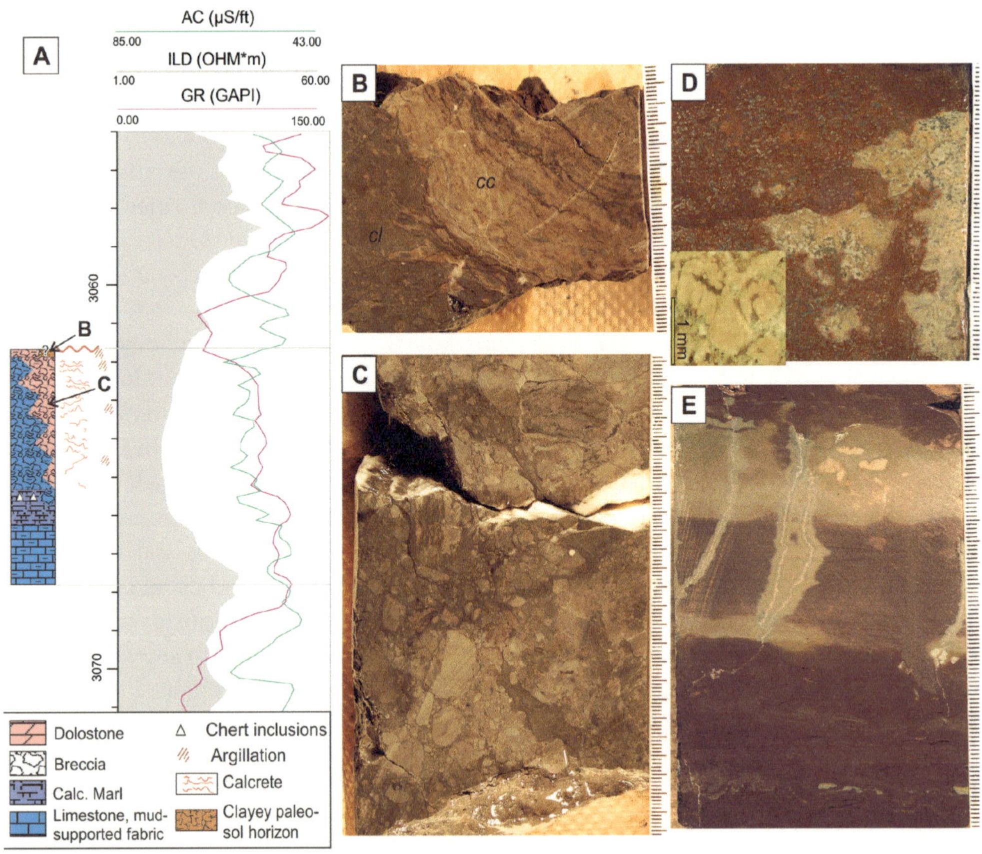

Figure 4. Permian disconformities of Sverdrup Basin in cores: (A–C) disconformity with a thick paleosol breccia at 3061.3 m, Graham C-52 well; (A) lithology and matched borehole logs; (B) core face photo of laminar calcrete crust (cc) interfingering with a claystone of probably upper paleosol horizon (cl); (C) calcretized and argillated breccia; locations of B and C are indicated on lithology. (D) Red mottled calcareous mudrock, Upper Pennsylvanian or basal Permian, Jameson Bay C-31, 2406.70 m; note dense Microcodium penetrations (mm-scale features); the inset shows typical Microcodium aggregates zoom with binocular microscope. (E) Red mottled bioturbated shale and siltstone, same age, Depot Island C-44, 1662–1663 m.

UWI	Well short name	TOP_ DEPTH (m)	BOT_ DEPTH (m)	Run (m)	Recovery (m)	Meteoric alteration (Y/N)	Unconformity surface present in core (Y/N)	TOP_ FORMATION	TOP_AGE	BOTTOM_ FORMATION	BOTTOM_AGE
300J607620110000	HECLA J-60	3603.7	3616.5	12.8	12.8	Y	Y	VAN HAUEN FM	GUADALUPIAN	VAN HAUEN FM	GUADALUPIAN
300C447630114000	DEPOT ISLAND C-44	1656.9	1675.2	18.3	17.4	Y	Y	CANYON FIORD FM	EARLY CISURALIAN	CANYON FIORD FM	EARLY CISURALIAN
300C447630114000	DEPOT ISLAND C-44	2457	2465.5	8.5	8.5	Y	Y	CANYON FIORD FM	PENNSYLVANIAN	IBBETT BAY FM	LOWER DEVONIAN
300L467630115000	SANDY POINT L-46	1786.1	1795.3	9.2	9.1	Y	N	CANYON FIORD FM	PENNSYLVANIAN	CANYON FIORD FM	PENNSYLVANIAN
300L467630115000	SANDY POINT L-46	1985.2	1988.5	3.3	3.3	Y	N	CANYON FIORD FM	PENNSYLVANIAN	CANYON FIORD FM	PENNSYLVANIAN
300C317650116300	JAMESON BAY C-31	2404	2413.1	9.1	9.1	Y	?	CANYON FIORD FM	PENNSYLVANIAN	CANYON FIORD FM	PENNSYLVANIAN
300F687720116300	SATELLITE F-68	2170.2	2173.5	3.3	3.3	Y	Y	TROLD FIORD FM	LATE GUADALUPIAN	TROLD FIORD FM	LATE GUADALUPIAN
300C527730090300	GRAHAM C-52	3061.7	3067.8	6.1	6.1	Y	N	HARE FIORD/ BELCHER CHANNEL	?SAKMARIAN (E. CISURALIAN)	HARE FIORD/ BELCHER CHANNEL	ASSELIAN (E. CISURALIAN)

Table 1. Cores with subaerial exposure profiles from the Upper Paleozoic of Sverdrup Basin, based on Ref. [91].

The thickest paleosol was encountered at 3061.3 m of Graham C-52 well (**Figure 4A–C**). The weak low-GR excursion just above the core top may record an upper clayey horizon of this paleosol or a transgressive deposit. This subaerial exposure profile may correlate to the unconformable contact of the Upper Nansen and Raanes formations (Asselian-Sakmarian boundary) of the basin margin zone [90]. Other disconformities from this core inventory occur stray within the defined third-order sequences, but may be assigned to higher frequency sea level fluctuations, as in the case of 9.1-m-thick core from the Belcher Channel Formation (lower Cisuralian) of Jameson Bay C-31 well described by Beauchamp et al. [90]. As stated in [90], these thinner (meter-scale) sequences or cyclothems are quite numerous in the Pennsylvanian—Lower Cisuralian (over 100 counted) but cannot be correlated between sections. Similarly, thin sequences in the Guadalupian part of the succession were traced based on well logs [84], but it is impossible to confirm subaerial nature of alleged sequence boundaries as no cores are available.

As the scanty core coverage in old exploration wells would not offer a chance to capture all stratigraphically meaningful disconformities, it is important to identify zones flushed by meteoric waters percolated from overlying subaerial surfaces. For example, in zones of meteoric oxidation, iron releases from decomposing synsedimentary sulfides and reprecipitates as ferric oxides and hydroxides. Seasonal waterlogging causes patchy reduction of iron into gley, and wetting-drying cycles usually imprint in characteristic red-gley mottling. Occurrence of oxidized basinal shales and siltstones with such mottling (**Figure 4D, E**) indicates fairly deep base level falls consistent with glacio-eustasy of the Late Paleozoic ice age [92]. Another feature indicative of paleovadose environment is *Microcodium* (**Figure 4D**), an aerobic microbially induced fabric abundant in Ca-rich subterranean environments of Pennsylvanian-Permian and late Cretaceous-Tertiary times but with no confirmed presence in rocks of other ages (**Figure 4D**; [93]).

6.2. Carboniferous of Moscow Basin, Russia

The Middle-Upper Mississippian and Pennsylvanian strata of the Moscow epicontinental basin of the central East European Craton (EEC) contain two cyclothemic successions dominated by shallow-marine carbonates and separated by a major Mississippian/Pennsylvanian unconformity [94]. The Upper Mississippian is a type succession for the Serpukhovian Stage, and Pennsylvanian strata host historical type sections for the Moscovian, Kasimovian, and Gzhelian stages of the Geological Time Scale [95]. The Mississippian/Pennsylvanian diastem (MPD) accounts for at least 10 My of late Serpukhovian-Bashkirian lowstand during which thick paleosols and deep (>110 m) fluvial incisions formed. Sequence stratigraphy of the two successions was developed based primarily on outcrops and disconformities which were used as main correlative horizons [96–99].

6.2.1. Middle-Upper Pennsylvanian

Similar to coeval classical cyclothems of North America [19, 86], Middle Pennsylvanian strata of the Moscow Basin have recorded a forced sea level control with drowning of seafloor into subphotic basinal environment on peaks of highstands and deep base level falls leading to subaerial exposure. These lowstands are thought to be the far-field response to expansion and shrinking of the Late Paleozoic ice dominantly from the Gondwanan icesheet [92]. Disconformities are scoured by transgressive erosion to various degrees, but some are onlapped with quiet-water facies with negligible truncation, preserving delicate features of former solums ("topclays"; **Figure 5A**). Fluvial facies or incised valleys are unknown. Topclays are palygorskitic, in some areas sepiolitic, indicating arid, well-drained pedogenic environments. A shift to montmorillonitic-illitic toplays recorded in the upper part of the studied succession flags the transition to slightly more humid climate. Other features are rare although deeply penetrating rhizoliths, shallow soil carbonate, low alumina/bases and Ba/Sr ratios, enhanced Mn and Sr, presence of soil gypsum and opal, and a characteristic peak in magnetic susceptibility, all suggesting a semiarid to arid pedogenic environment. The palygorskite clay of this paleo-pedon retains 1.1–1.5% of connate organic matter which is fulvate-dominated resembling organic matter from present-day aridisols [100]. The succession seems to be flushed throughout with meteoric fluids and repeatedly exposed to vadose environment, which left the penetrative systems of small solution vugs and oxidized organic matter and pyrites in basinal and siliciclastics-rich units.

6.2.2. *Middle-Upper Mississippian*

This ~90-m-thick shallow-marine succession deposited during Late Viséan and lower-middle Serpukhovian (~16 My) is composed of shallow-marine limestone-dominated units bounded by six main disconformities and even more weakly developed subaerial surfaces that could not be traced between outcrops [99]. Fluvial and deltaic floodplain siliciclastics wedge between Viséan limestone units from southwest. The Viséan strata show a number of unusual sedimentary features, such as a lack of high-energy facies, shallow-subtidal marine sediments penetrated by *Stigmaria,* and beds of palustrine marls (*sensu* [101]) composed of a mixture of authigenic saponite, beidellite, and micritic calcite with strong negative offset of $\delta^{13}C$. Disconformities range in expression from undercoal solution-collapse horizons of only a few cm thick to deep paleokarsts. Incised fluvial channels are reported at two stratigraphic levels to the west and north of the study area. The deepest incisions (>15 m) developed from the Kholm disconformity, and this stratigraphic break is also marked with the deepest paleokarst profile (**Figure 4B**). All paleosol profiles contain evidence of rooting activity with numerous *Stigmaria* (rooting systems of arborescent lycopsids). The uppermost studied paleosol below

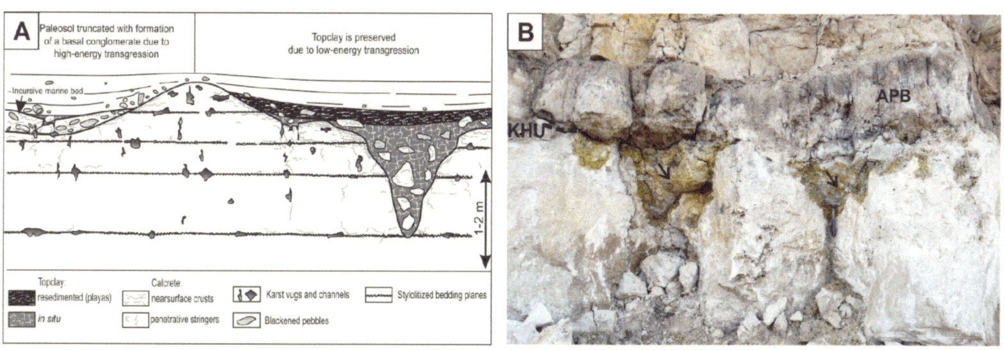

Figure 5. Paleosols and Paleokarsts at Carboniferous disconformities of Moscow Basin: (A) major elements and variability of upper Middle Pennsylvanian disconformities of Moscow Basin, slightly modified from Kabanov et al. [27]; (B) Kholm disconformity in top of Mikhailovian (KHU) and Akulshino palustrine marl (APB) at Novogurovsky Quarry, slightly modified from Kabanov et al. [99]; yellow clayey paleosol in solution pockets is arrowed.

the MPD is mid-Serpukhovian in age. It is a thin palygorskitic calcrete [99] formed under significantly drier climate than underlying *Stigmaria*-bearing paleosols. Paleosol mineralogy and proxies for pedogenic environments are discussed in [102, 103].

6.2.3. Disconformities in cores

Paleosols and karstified profiles of Middle-Late Mississippian and Pennsylvanian age are frequently intersected by cores in oil and gas exploration areas of the eastern EEC (**Figure 6A–C**). Project geologists usually ignore these surfaces. However, eroded disconformities invisible with geophysical tools may record prolonged hiatuses, as indicated by thick rhizocretions left by perennial plants requiring fairly thick soil cover to root in (**Figure 6C**).

6.3. Lower-Middle Devonian of Mackenzie Corridor, Northwestern Canada

Devonian strata of the central and northern Mackenzie Corridor located within the limits of ancestral North America are composed of Lower Devonian-Eifelian shallow-marine carbonates, dolostone breccias, and evaporites; Givetian-Frasnian basinal shales of the Horn River Group hosting isolated carbonate platforms (banks) of Ramparts Formation; and the Frasnian-Famennian Imperial Formation composed of fine-grained turbiditic siliciclastics and coarse-grained siliciclastics and chert conglomerates of the Tuttle Formation. The latter straddles the Devonian-Carboniferous boundary (**Figure 7**; [40, 104]).

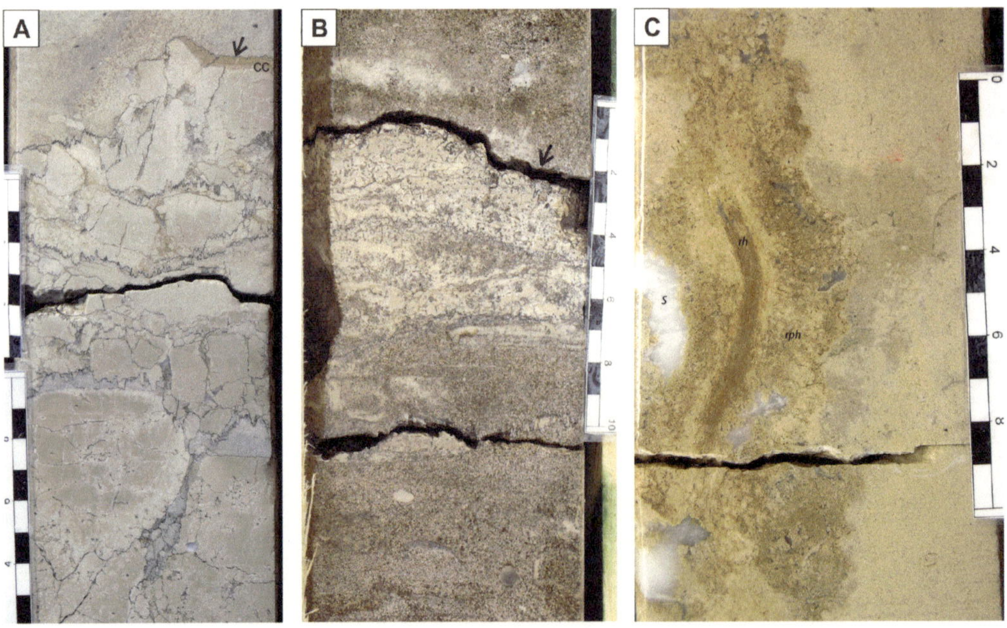

Figure 6. Eroded paleosols on core face of shallow-marine limestones, Bashkirian (Lower Pennsylvanian), southeastern EEC: (A) collapsed karst breccia with thin laminar calcrete crusts (cc); (B) more massive calcrete crust with rootlet channels; ravinement surface is arrowed; (C) rhizolith (*rh*) with thick peripheral alteration zone (*rph*) found in 3.5 m below a disconformity; (s) is anhydrite fill of karst voids; scale bar in centimeters.

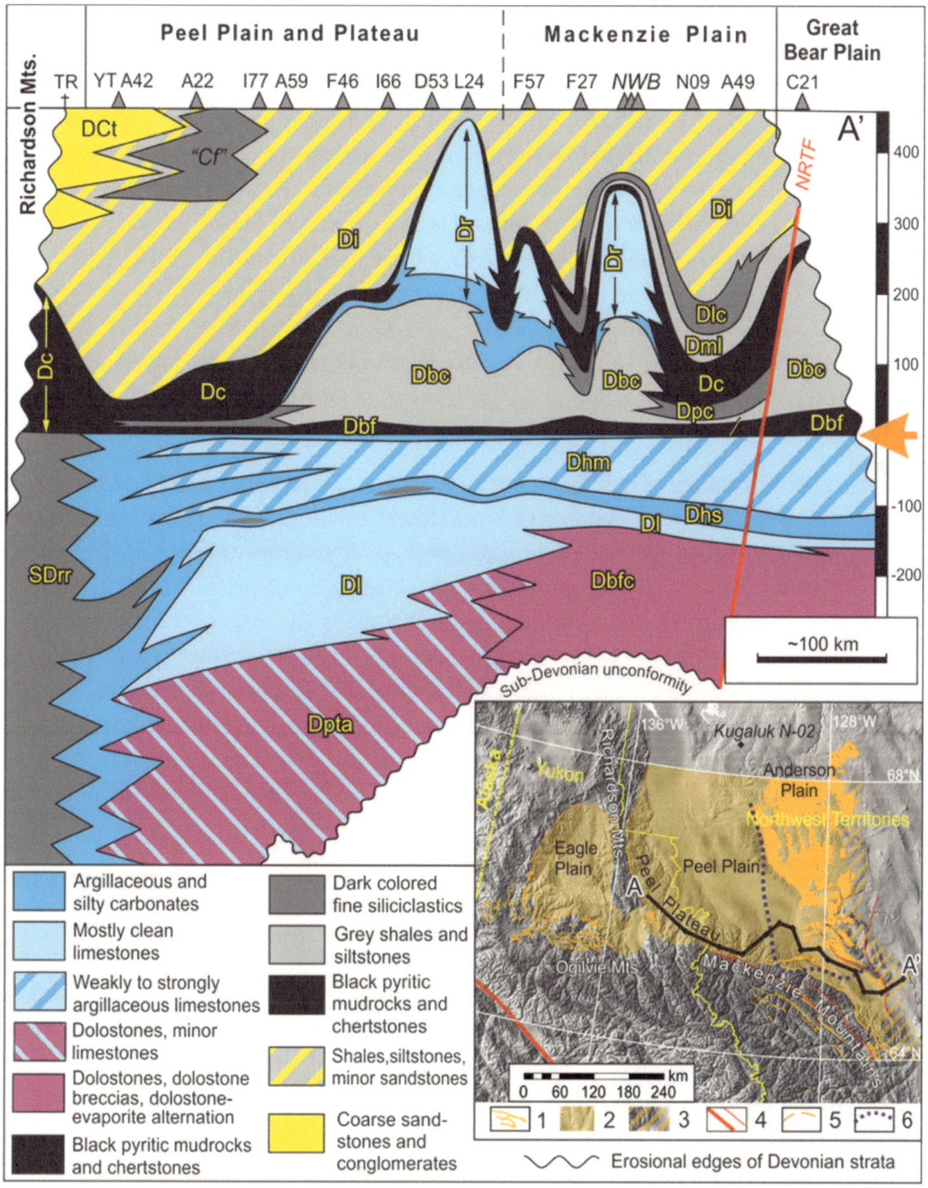

Figure 7. Devonian succession of central and northern Mackenzie Corridor on a cross-section A-A′ anchored on mid-Devonian drowning unconformity (arrowed). TR is Trail River outcrop section; wells from left to right: Cranswick YT A-42, Cranswick A-22, S. Ramparts I-77, N. Ramparts A-59, Ramparts River F-46, Hume River I-66, Hume River D-53, Carcajou L-24, Maida Creek F-57, Hoosier F-27, NWB is Norman Wells oilfield, Little Bear N-09, Bluefish A-49, and Bracket Lake C-21. Stratigraphic units: (Dbfc) Bear Rock, Fort Norman, and Camsell fms.; (Dpta) Peel, Tatsieta, and Arnica formations; (Dl) Landry Fm.; (SDrr) Road River Group; (Dhs) Headless Mbr. of Hume Fm; (Dhm) Hare Indian Fm.; (Dbf) Bluefish Mbr.; (Dbc) Bell Creek Mbr.; (Dfc) Francis Creek Mbr.; (Dpc) Prohibition Creek Mbr.; (Dr) Ramparts Fm.; (Dc) Canol Fm.; Imperial Fm. undivided (Di); (Dml) Mirror Lake Mbr.; (Dlc) Loon Creek Mbr.; (DCt) Tuttle Fm.; ("Cf") informal unit *Cf*. Inset map shows wide occurrence of the Horn River Group between 64 and 68 parallel in (1) outcrops, (2) subsurface, and (3) patchy presence in erosional outliers; (4) Tintina Fault Zone (thick) and smaller scale main faults in the Mackenzie Foldbelt (thin); (5) Canol Formation dips beneath thick siliciclastic wedge of Imperial and Tuttle formations; (6) paleogeographic offshore limits of thick Hare Indian siliciclastics (Bell Creek Mbr.) and overlying Ramparts Limestone. The eastern limit of Laramide deformation front is approximated by Norman Range thrust fault (NRTF).

A shallow-marine peritidal succession of Emsian age measured in the nearly continuous core of Kugaluk N-02 well (**Figure 7**) contains 86 disconformities that bear distinct signatures of subaerial exposure (rank 0, 1, and 2 discontinuities in **Figure 8**). Of these, 43 surfaces are marked with thick (>1 m) paleokarst profiles and 3 surfaces by thick rubbly paleosols and several meters of karstified rock below [105, 106]. This 440 m thick succession deposited over a period of 15–18 My, assuming that the top of Landry Formation approximates to the base of Eifelian [105, 107] and Delorme/Arnica contact is found in the Lochkovian or Pragian [108]. However, only seven subaerial exposure profiles have been identified in the Arnica—lower Landry part of this succession in the outcrop section measured at Rumbly Creek West Ridge, including one deep profile with thick paleosol [109]. Given very similar shallow-water facies assemblage of this outcrop and Kugaluk N-02 core, small number of disconformities appears to be an artifact of poor preservation of the weathered section and limited time spent on it by the examiner.

Subaerial exposure profiles of similar character are very common in Lower and basal Middle Devonian cores over the broad expanse of Mackenzie Corridor. Some thick profiles show signature of prolonged exposure and multiphase pedogenic overprinting resulted in complete loss of sedimentary fabrics, as exemplified by a mature paleosol profile at 600.25–603.5 m of Ebbutt D-50 well (**Figure 9B–D**). One interesting feature is the absence of root penetrations that are characteristic of younger Phanerozoic paleosols (**Figure 6**), which is interpreted as an evolutionary imprint of prevascular plant landscape. Small (<1 mm in diameter) rhizocretions occur only in thin marshland beds (palustrine facies; [101]) occupying incursive and transgressive positions in peritidal sequences of Landry Formation [99]. This "palustrine facies" has been also identified in outcrop [109]. Like in described above Late Paleozoic examples, none of available geophysical logs can be relied upon to trace even thickest paleosols of this type in the subsurface (**Figure 9**).

6.4. Mid-Devonian drowning unconformity of Mackenzie Corridor

Bioturbated and richly fossiliferous benthic limestones of Hume Formation containing a diverse benthic fauna are onlapped by black calcareous laminated shales of the Bluefish Member. The onlap surface is a strong seismic reflector commonly used as stratigraphic datum (**Figure 7**). In the project area (**Figure 7**), the surface appears table flat on outcrop scale, if not tectonically displaced, but in the southern Mackenzie Corridor it is outgrown by pinnacle-shaped carbonate buildups referred to as Horn Plateau reefs [110, 111].

The Hume/Bluefish contact has been measured in three cores from Canol Shale exploration wells and accessed in three outcrops of the Norman Range and northern Mackenzie Mountains [45, 109]. The coral-stromatoporoid facies composing the main part of the upper Hume Formation occurs in direct contact with the Bluefish shale in two of six sections, and in both cases, it shows a rugged corroded top with deep (8 cm in core) solution pockets filled with black shale from the overlying anoxic facies. The upper few decimeters below the top are chertified and also very pyritic in core or rusty in outcrops. Phosphatic crusts

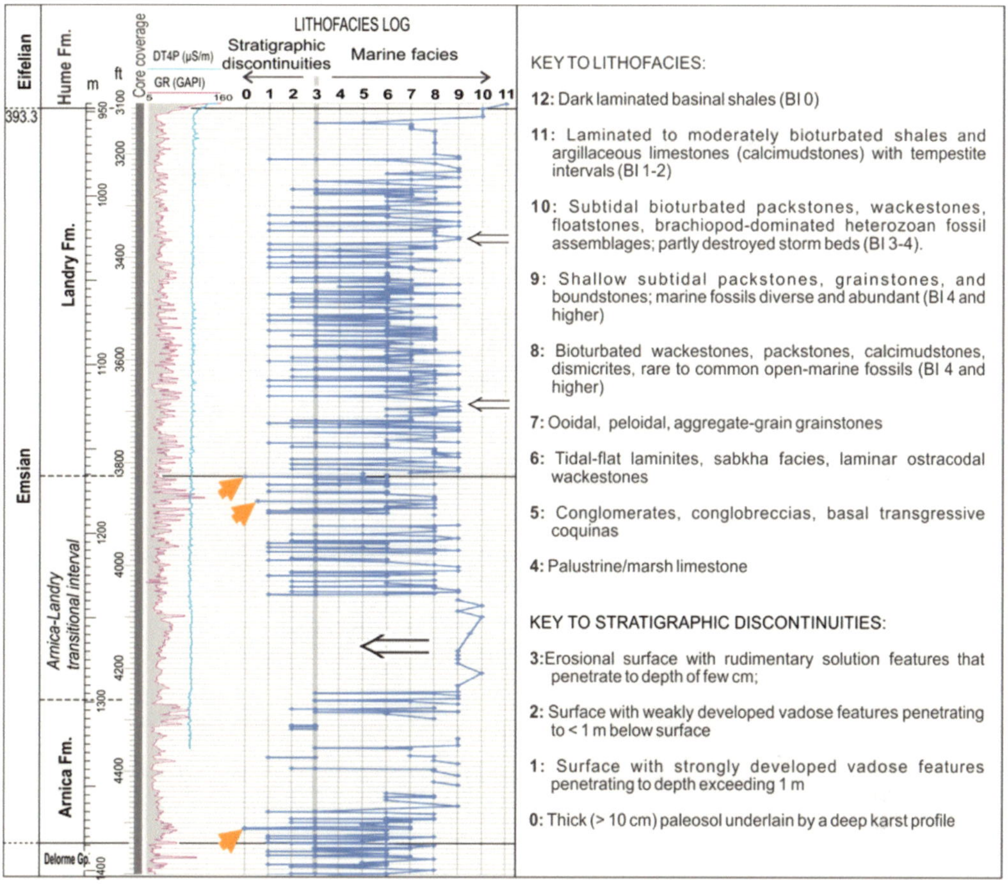

Figure 8. Lithofacies log for the Arnica-Landry succession in core of Kugaluk N-02 well with ranked disconformities, modified from Kabanov [105, 106]. Each facies point represents a mid-point of the descriptive interval. "No information" gaps in joint line indicate dolostones with obliterated sedimentary fabrics or "lost cores" from fractured zones. Black hollow arrows point at thick highstand intervals with offshore lithofacies and no disconformities. Orange arrows point at thickest subaerial exposure profiles with preserved paleosols.

characteristic of hardgrounds at other drowning unconformities did not develop, which is explained by overall phosphorus-lean sedimentary system [45]. Four other sections show 0.5–2.6-m-thick transitional interval of argillaceous bioturbated micritic limestones and shales. This transitional interval contains smooth discontinuity surfaces but no rugged hardgrounds. This transitional limestone contains brachiopod banks but no stromatoporoids. Pelagic tentaculitids appear in this unit and become rock-forming in base of Bluefish Member. The top of this transitional unit is usually smooth and probably storm-scoured. The basal few cm of the Bluefish Member characteristically contain lag concentrate of imbricated brachiopod shells mixed with diverse tentaculitids, sometimes dominated by tentaculitids with rare disintegrated brachiopod valves. Bioturbation in this basal Bluefish

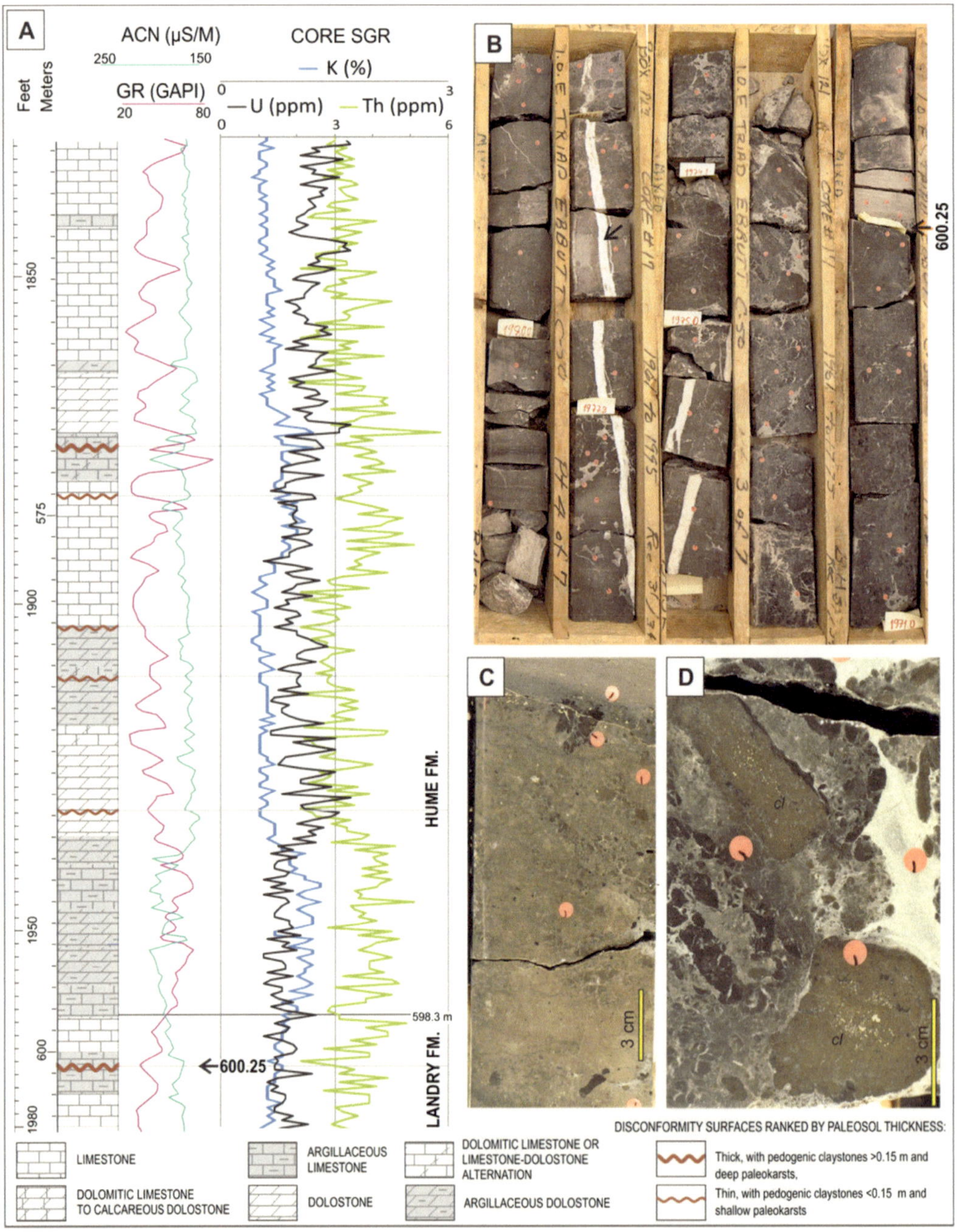

Figure 9. A shallow-marine peritidal succession measured in core of Ebbutt D-50 well (southern Mackenzie Corridor): (A) Striplog showing lack of well log response at multiple paleosols. (B–D) Polished and etched core face with details of paleosol at 600.25 m; (B) box view with pedogenic claystone-to-calcrete at 600.25–602.0 m (1969.3–1975.0 ft) and intense alteration down to at least 603.5 m (1980.0 ft); black arrow points at hydrothermal dolostone vein. (C) Top of paleosol profile composed of multiphase clayey calcrete; (D) float breccia at 600.9 m with residual clasts of marine limestone (*cl*). Sticky marks are ED-XRF reading points.

bed drops abruptly to BI ≈ 2 and right above this bed declines to zero. Enrichment in chalcophyle trace metals in enrichment factor notation (EFV and EFMo) grows gradually from moderate in the base of the Bluefish Member to a traceable spike of high values in 2.0 m above the base, indicating a gradual spread of anoxia.

7. Conclusion: a word of caution

It is generally accepted that the unconformities collectively record about 90% of the geologic time in its stratal expression. Stratigraphic unconformities are critical surfaces in sequence stratigraphy, but their identification remains largely the art of a visual rock assessment. Subaerial exposure profiles with paleosols are most common expression of non-eroded disconformities, but gamma and other conventional log signatures of even thick pedogenic claystones tend to stay at the background of host strata, and the majority of these surfaces do not coincide with surfaces of lithological change that would produce impedance contract for a seismic survey. Although major surfaces with prominent paleokarsts, erosion relief, lateritic mantles, and/or system of incised channels are certainly correlated, it has to be admitted that straightforward and universal technique to identify disconformities in coreless subsurface sections does not exist.

Stratigraphic unconformities included in table of formations are usually biased to those surfaces that were identified in outcrop, and their correlation may be undermined by a blank zone of unknown surfaces below and above, especially when dealing with non-cored intervals in the subsurface. This bias improves with increasing knowledge on the stacking pattern and ranking of measured disconformities.

Stratigraphic breaks diagnosed in old times and supported by missing faunal zones (e.g., sub-Canol hiatus of Mackenzie Corridor) are prone to dissolution or narrowing with increasing accuracy of biostratigraphic framework and absolute dating. Robustness of identified hiatuses should be confirmed with signatures of subaerial exposure or erosion.

Drowning unconformities are drowning surfaces specific for carbonate platforms. Usually, such surfaces produce vivid reflection horizons, and in the subsurface, they frequently have better stratigraphic value than platform-embedded subaerial disconformities.

Acknowledgements

This work is a contribution to the Mackenzie and Western Arctic projects of the Geomapping for Energy and Minerals Program (GEM-2) and IGCP-652 "Reading geologic time in Paleozoic sedimentary rocks: the need for an integrated stratigraphy." Cordial thanks are due to the volume editor and reviewer Gemma Aiello, GSC reviewer Keith Dewing, and my better half Darya Baranova who worked through text and literature formatting. Colleagues of GSC Milovan Fustic and John W. Shimeld are greatly thanked for contributing figures. This is the NRCan Lands and Minerals Sector contribution # 20170095.

Author details

Pavel Kabanov

Address all correspondence to: pavel.kabanov@canada.ca

Geological Survey of Canada, Natural Resources Canada, Calgary, Canada

References

[1] Miall AD. The valuation of unconformities. Earth-Science Reviews. 2016;**163**:22-71

[2] Dunbar CO, Rodgers J. Principles of Stratigraphy. New York: Wiley & Sons; 1957. p. 356

[3] Hutton J. Theory of the Earth, with Proofs and Illustrations. Vols. 1 and 2. Edinburgh: Creech JW; 1795 (facsimile edition, Codicote: Wheldon & Wesley; 1959)

[4] Bakewell R. Introduction to Geology. 2nd ed. London: R. Taylor; 1815. 362 p

[5] Le Conte J. Elements of Geology. 3rd ed. New York: D. Appleton & Co., 1893

[6] Darwin CR. On the Origin of Species by Means of Natural Selection or the Preservation of Favoured Races in the Struggle for Life. London: Murray; 1859

[7] Miall AD. Stratigraphy: A Modern Synthesis. Berlin: Springer-Verlag; 2016. 454 p

[8] Blackwelder E. The valuation of unconformities. Journal of Geology. 1909;**17**:289-299

[9] Grabau AW. Principles of Stratigraphy. New York: A. G. Seiler and Company; 1913. 1150 p

[10] Pirsson LV, Schuchert C. A Text-book of Geology. New York: Wiley; 1915. 1092 p

[11] Barrell J. Rhythms and the measurement of geologic time. Geological Society of America Bulletin. 1917;**28**:745-904

[12] Sloss LL, Krumbein WC, Dapples EC. Integrated facies analysis. In: Longwell CR, editor. Sedimentary Facies in Geologic History. Vol. 39. Geological Society of America Memoir. Whitefish, MT: Literary Licensing, LLC, 1949. pp. 91-124

[13] Sloss LL. Sequences in the cratonic interior of North America. Geological Society of America Bulletin. 1963;**74**:93-114

[14] Mitchum R, Vail P, Thompson III S. Seismic stratigraphy and global changes in sea level, part 2: The depositional sequence as the basic unit for stratigraphic analysis. In: Payton CE, editor. Seismic Stratigraphy: Application to Hydrocarbon Exploration. Vol. 26. AAPG Memoir. Tulsa, OK: AAPG,1977. pp. 53-62

[15] Vail PR, Mitchum RM Jr, Thompson III S. Seismic stratigraphy and global changes of sea level, part 3: Relative changes of sea level from coastal onlap. In: Payton CE, editor.

Seismic Stratigraphy: Applications to Hydrocarbon Exploration. Vol. 26. AAPG Memoir. Tulsa, OK: AAPG, 1977. pp. 63-81

[16] Galloway WE. Genetic stratigraphic sequences in basin analysis, I. Architecture and genesis of flooding-surface bounded depositional units. American Association of Petroleum Geologists Bulletin. 1989;**73**:125-142

[17] Posamentier HW, Allen GP. Siliciclastic sequence stratigraphy: Concepts and applications. Vol. 7. SEPM Concepts in Sedimentology and Paleontology. 1999. 210 p

[18] Schlager W. Type 3 sequence boundaries. In: Harris PM, Saller AH, Simo JA, editors. Advances in Carbonate Sequence Stratigraphy: Application to Reservoirs, Outcrops, and Models. Vol. 63. Society of Economic Paleontologists and Mineralogists, Special Publication. Tulsa, OK: SEPM, 1999. pp. 35-46

[19] Heckel PH. Evaluation of evidence for glacio-eustatic control over marine Pennsylvanian cyclothems in North America and consideration of possible tectonic effects. In: Dennison JM, Ettensohn FR, editors. Tectonic and Eustatic Controls on Sedimentary Cycles. Vol. 4. SEPM Concepts in Sedimentology and Paleontology. Tulsa, OK: SEPM, 1994. pp. 65-87

[20] Catuneanu O. Principles of Sequence Stratigraphy. Amsterdam, Boston, Heidelberg: Elsevier; 2006

[21] Catuneanu O, Abreu V, Bhattacharya JP, Blum MD, Dalrymple RW, Eriksson PG, Fielding CR, Fisher WL, Galloway WE, Gibling MR, Giles KA, Holbrook JM, Jordan R, Kendall CGStC, Macurda B, Martinsen OJ, Miall AD, Neal JE, Nummedal D, Pomar L, Posamentier HW, Pratt BR, Sarg JF, Shanley KW, Steel RJ, Strasser A, Tucker ME, Winker C. Towards the standardization of sequence stratigraphy. Earth-Science Reviews. 2009;**92**:1-33

[22] Embry AF. Transgressive-regressive (TR) sequence analysis of the Jurassic succession of the Sverdrup Basin, Canadian Arctic Archipelago. Canadian Journal of Earth Sciences. 1993;**30**:301-320

[23] Embry AF. Practical Sequence Stratigraphy. Canadian Society of Petroleum Geologists. Calgary, AB, 2009. p. 1-81. [Internet] Available from: http://www.cspg.org

[24] Embry AF. Correlating siliciclastic successions with sequence stratigraphy. In: Application of Modern Stratigraphic Techniques: Theory and Case Histories. SEPM Special Publication. Vol. 94. Tulsa, OK: SEPM, 2010. pp. 35-53

[25] Veeken PCH. Seismic stratigraphy, basin analysis and reservoir characterization. Seismic Exploration. 2007;**37**:489

[26] Blum MD, Törnqvist TE. Fluvian responses to climate and sea-level change: A review and look forward. Sedimentology. 2000;**47**(supplement 1):2-48

[27] Kabanov P, Alekseeva T, Alekseev A, Alekseeva V, Gubin S. Paleosols in Late Moscovian (Carboniferous) marine carbonates of East European Craton revealing "Great Calcimagnesian Plain" paleolandscapes. Journal of Sedimentary Research. 2010;**80**:195-215

[28] Martinius AW, Fustic M, Garner DL, Jablonski BVJ, Strobl RS, MacEachern JA, Dashtgard SE. Reservoir characterization and multiscale heterogeneity modeling of inclined heterolithic strata for bitumen-production forecasting, McMurray Formation, Corner, Alberta, Canada. Marine and Petroleum Geology. 2017;**82**:336-361

[29] Deptuck ME, MacRae RA, Shimeld JW, Williams GL, Fensome RA. Revised Upper Cretaceous and lower Paleogene lithostratigraphy and depositional history of the Jeanne d'Arc Basin, offshore Newfoundland, Canada. AAPG Bulletin. 2003;**87**:1459-1483

[30] Christie-Blick N, Mountain GS, Miller KG. Seismic stratigraphic record of sea-level change. In: Sea-level Change. National Academy of Sciences, Studies in Geophysics; Washington, D.C.,1990. pp. 116-140

[31] Schlager W. Carbonate Sedimentology and Sequence Stratigraphy. Vol. 8. SEPM Concepts in Sedimentology and Paleontology. Tulsa, OK: SEPM, 2005. 200p

[32] Twidale CR. River patterns and their meaning. Earth-Science Reviews. 2004;**67**:159-218

[33] Miller CR, James NP, Bone Y. Prolonged carbonate diagenesis under an evolving late Cenozoic climate: Nullarbor Plain, southern Australia. Sedimentary Geology. 2012;**261-262**:33-49

[34] Mindszenty A, D'Argenio B, Aiello G. Lithospheric bulges recorded by regional unconformities: The case of Mesozoic-Tertiary Apulia. Tectonophysics. 1995;**252**:137-161

[35] Kaufmann B. Calibrating the Devonian time scale: A synthesis of U–Pb ID-TIMS ages and conodont stratigraphy. Earth Science Reviews. 2006;**76**:175-190

[36] Powell MG. Climatic basis for sluggish macroevolution during the late Paleozoic ice age. Geology. 2005;**33**:381-381

[37] Groves JR, Yue W. Foraminiferal diversification during the late Paleozoic ice age. Paleobiology. 2009;**35**:367-392

[38] Hume GS, Link TA. Geological investigations in the Mackenzie River area, Northwest Territories. Geological Survey of Canada. 1945;**45-16**:1-87

[39] Braun WK, Norris AW, Uyeno TT. Late Givetian to early Frasnian biostratigraphy of Western Canada: The Slave Point-Waterways boundary and related events. In: McMillan NJ, Embry AF, Glass DJ, editors. Devonian of the World. Vol. 14-3. Canadian Society of Petroleum Geologists. Calgary, AB: ISPG, 1989. pp. 93-111

[40] Norris AW. Chapter 7: Devonian. In: Norris DK, editor. Geology and Mineral and Hydrocarbon Potential of Northern Yukon Territory and Northwestern District of Mackenzie. Vol. 422. Geological Survey of Canada Bulletin. 1997. pp. 163-200

[41] Muir ID. Devonian Hare Indian and Ramparts formations, Mackenzie Mountains, N.W.T.: Basin-fill, platform and reef development [thesis]. Ottawa: University of Ontario; 1988

[42] Pugh DC. Pre-Mesozoic geology in the subsurface of Peel River Map area, Yukon Territory and district of Mackenzie. Geological Survey of Canada. 1983;**401**:1-61

[43] Pyle LJ, Gal LP. Reference section for the Horn River Group and definition of the Bell Creek Member, Hare Indian Formation in central Northwest Territories. Bulletin of Canadian Petroleum Geology. 2016;**64**:67-98

[44] Rocheleau J, Fiess KM. Northwest Territories oil and gas poster series: Basins and Petroleum resources, table of formations, schematic cross sections. NWT Open File Report. 2014;3; 2 posters

[45] Kabanov P, Gouwy SA. The Devonian Horn River Group and the basal Imperial Formation of the central Mackenzie Plain, N.W.T., Canada: Multiproxy stratigraphic framework of a black-shale basin. Canadian Journal of Earth Sciences. 2017;**54**:409-429

[46] Tassonyi EJ. Subsurface geology, lower Mackenzie River and Anderson River area, District of Mackenzie. Geological Survey of Canada Paper;1969;**68-25**:1-207

[47] Retallack GJ. Soils of the Past: An Introduction to Paleopedology. 2nd ed. Blackwell, printed in Oxford and Northampton (UK), Alden Press Ltd; 2001. 404 p

[48] Wright VP. Paleosols in shallow marine carbonate sequences. Earth-Science Reviews. 1994;**35**:367-395

[49] Wang ZS, Rasbury ET, Hanson GN, Meyers WJ. Using the U-Pb system of calcretes to date the time of sedimentation of clastic sedimentary rocks. Geochimica et Cosmochimica Acta. 1998;**62**:2823-2835

[50] Rasbury ET, Cole JM. Directly dating geological events: U-Pb dating of carbonates. Reviews of Geophysics. 2009;**47**:1-27

[51] Tabor NJ, Myers TS. Paleosols as indicators of paleoenvironment and paleoclimate. Annual Review of Earth and Planetary Sciences. 2015;**43**:333-361

[52] Allen GH, Pavelsky TM. Patterns of river width and surface area revealed by the satellite-derived North American River Width data set. Geophysical Research Letters. 2015;**42**:395-402

[53] Downing JA, Prairie YT, Cole JJ, Duarte CM, Tranvik LJ, Striegl RG, McDowell WH, Kortelainen P, Caraco NF, Melack JM, Middelburg JJ. Global abundance and size distribution of lakes, ponds, and impoundments. Limnology and Oceanography. 2012;**51**:2388-2397

[54] Kraus MJ. Paleosols in clastic sedimentary rocks: Their geologic applications. Earth Science Reviews. 1999;**47**:41-70

[55] Webster T. Observations on the Purbeck and Portland beds. Geological Society of London Transactions. 1826;**2**:37-44

[56] Buckland W. Geology and Mineralogy Considered with Reference to Natural Theology. Vols. 1 and 2. London: W. Pickering; 1837

[57] Retallack GJ. A short history and long future for paleopedology: New Frontiers in Paleopedology and Terrestrial Paleoclimatology. SEPM Special Publication 2013;**104**:5-16

[58] Polynov BB. Contributions of Russian Scientists to Paleopedology. Leningrad: USSR Academy of Sciences; 1927. 32 p

[59] Buol SW, Hole FD, McCracken RW. Soil Genesis and Classification. 4th ed. Ames, IA: Iowa State University Press, 1997

[60] Nikiforoff CC. Reappraisal of the soil. Science. 1959;**129**:186-196

[61] Merrill GP. Rocks, Rock-weathering and Soils. New York: MacMillan Company; 1897. 411 p

[62] Taylor G, Eggleton RA. Regolith Geology and Geomorphology. New York: Wiley; 2001. 384 p

[63] Longman MW. Carbonate diagenetic textures from nearshore diagenetic environments. Bulletin American Association of Petroleum Geology. 1980;**64**:461-487

[64] National Research Council (NRC). Basic Research Opportunities in Earth Science. Washington, DC: National Academy Press; 2001

[65] Lin H. Earth's Critical Zone and hydropedology: Concepts, characteristics, and advances. Hydrology and Earth System Sciences. 2010;**14**:24-45

[66] Soil Survey Staff. Keys to Soil Taxonomy. Washington, DC: Nat. Resour. Conserv. Serv.; 2010

[67] Sheldon ND, Tabor NJ. Quantitative paleoenvironmental and paleoclimatic reconstruction using paleosols. Earth-Science Reviews. 2009;**95**:1-52

[68] Adams JS, Kraus MJ, Wing SL. Evaluating the use of weathering indices for determining mean annual precipitation in the ancient stratigraphic record. Palaeogeography, Palaeoclimatology, Palaeoecology. 2011;**309**:358-366

[69] Esteban M, Klappa CF. Subaerial exposure environments. In: Scholle PA, Bebout DG, Moore CH, editors. Carbonate Depositional Environments. American Association of Petroleum Geologists Memoir. Vol. 33; 1983. pp. 1-96

[70] Wright VP. Calcretes. In: Nash D, McLaren S, editors. Geochemical Sediments and Landscapes. Wiley, Chichester, West Sussex, UK, etc.: Blackwell; 2007. pp. 10-45

[71] Alonso-Zarza AM, Wright VP. Chapter 5. Calcretes. In: Alonso-Zarza AM, Tanner L, editors. Carbonates in Continental Settings: Facies, Environments, and Processes. Vol. 61. Developments in Sedimentology. Amsterdam, etc.: Elsevier, 2010. pp. 225-268

[72] Schlager W. Drowning unconformities on carbonate platforms. In: Crevello PD, Wilson JL, Sarg JF, Read JF, editors. Controls on Carbonate Platform and Basin Development. Vol. 41. Society of Economic Paleontologists and Mineralogists Special Publication. Tulsa, OK: SEPM, 1989. pp. 15-25

[73] Bosellini A, Morsilli M. A Lower Cretaceous drowning unconformity on the eastern flank of the Apulia Platform (Gargano Promontory, southern Italy). Cretaceous Research. 1997;**18**:51-61

[74] Godet A. Drowning unconformities: Palaeoenvironmental significance and involvement of global processes. Sedimentary Geology. 2013;**293**:45-66

[75] Zampetti V, Schlager W, van Konijnenburg J-H, Evert A-J. Architecture and growth history of a Miocene carbonate platform from 3D seismic reflection data; Luconia province, offshore Sarawak, Malaysia. Marine and Petroleum Geology. 2004;**21**:517-534

[76] Christie-Blick N. Onlap, offlap, and the origin of unconformity-bounded depositional sequences. Marine Geology. 1991;**97**:35-56

[77] Schlager W. Benthic carbonate factories of the Phanerozoic. International Journal of Earth Sciences. 2003;**92**:445-464

[78] Lachlear N, Guiraud M, El Harif A, Dommergues JL, Dera G, Durlet C. Early Jurassic normal faulting in a carbonate extensional basin: Characterization of tectonically driven platform drowning (High Atlas rift, Morocco). Journal of the Geological Society. 2009;**166**:413-430

[79] Lavoie D. Diachronous tectonic collapse of the Ordovician continental margin, eastern Canada: Comparison between the Quebec Reentrant and St. Lawrence Promontory. Canadian Journal of Earth Sciences. 1994;**31**:1309-1319

[80] Brett CE, Baird GC. Revised stratigraphy of the Trenton Group in its type area, central New York State: Sedimentology and tectonics of a Middle Ordovician shelf-to-basin succession. Physics and Chemistry of the Earth. 2002;**27**:231-263

[81] Minzoni M, Lehrmann DJ, Dezoeten E, Enos P, Montgomery P, Berry A, Qin A, Meiyi Y, Ellwood BB, Payne JL. Drowning of the Triassic Yangtze Platform, South China, by tectonic subsidence into toxic deep waters of an anoxic basin. Journal of Sedimentary Research. 2015;**85**:419-444

[82] Kidder DL, Worsley TR. Phanerozoic Large Igneous Provinces (LIPs), HEATT (Haline Euxinic Acidic Thermal Transgression) episodes, and mass extinctions. Palaeogeography, Palaeoclimatology, Palaeoecology. 2010;**295**:162-191

[83] Karstensen J, Stramma L, Visbeck M. Oxygen minimum zones in the eastern tropical Atlantic and Pacific oceans. Progress in Oceanography. 2008;**77**:331-350

[84] Jenkyns HC. Geochemistry of oceanic anoxic events. Geochemistry, Geophysics, Geosystems. 2010;11:Q03004. DOI: 10.1029/2009GC002788. 30 pp

[85] Föllmi KB, Gainon F. Demise of the northern Tethyan Urgonian carbonate platform and subsequent transition towards pelagic conditions: The sedimentary record of the Col de la Plaine Morte area, central Switzerland. Sedimentary Geology. 2008;**205**:142-159

[86] Algeo TJ, Heckel PH. The Late Pennsylvanian Midcontinent Sea of North America: A review. Palaeogeography, Palaeoclimatology, Palaeoecology. 2008;**268**:205-221

[87] Beauchamp B, Harrison JC, Henderson CM. Upper Paleozoic stratigraphy and basin analysis of the Sverdrup Basin, Canadian Arctic Archipelago. Part 2 – Transgressive-regressive sequences. In: Current Research, Part G: Geological Survey of Canada Paper. Vol. 89-1G. Calgary, AB: ISPG, 1989. pp. 115-124

[88] Beauchamp B, Henderson CM. The Lower Permian Raanes, Great Bear Cape, and Trappers Cove formations, Sverdrup Basin, Canadian Arctic: Stratigraphy and conodont zonation. Bulletin of Canadian Petroleum Geology. 1994;**42**:562-597

[89] Embry AF, Beauchamp B. Sverdrup Basin. In: Miall AD, editor. Sedimentary Basins of the World. Vol. 5. Amsterdam, etc.: Elsevier, 2008. pp. 451-471

[90] Beauchamp B, Harrison JC, Utting J, Brent TA, Pinard S. Carboniferous and Permian subsurface stratigraphy, Prince Patrick Island, Canadian Arctic. Geological Survey of Canada, Bulletin. 2001;**565**:83

[91] Kabanov PB, Dewing KE. Geological and geochemical data from the Canadian Arctic Islands. Part XII: Descriptions and lithologs of Upper Paleozoic core. Geological Survey of Canada Open File. 2014;**7569**:121

[92] Rygel MC, Fielding CR, Frank TD, Birgenheier LP. The magnitude of Late Paleozoic glacioeustatic fluctuations: A synthesis. Journal of Sedimentary Research. 2008;**78**:500-511

[93] Kabanov P, Anadon P, Krumbein WE. Microcodium: An extensive review and a proposed non-rhizogenic biologically induced origin for its formation. Sedimentary Geology. 2008;**205**:79-99

[94] Alekseev AS, Kononova LI, Nikishin AM. The Devonian and Carboniferous of the Moscow Syneclise (Russian Platform): Stratigraphy and sea-level changes. Tectonophysics. 1996;**268**:149-168

[95] Davydov VI, Schmitz M, Korn D. The Carboniferous Period. In: Gradstein F, Ogg J, Schmitz M, Ogg G., editors. The Geological Time Scale. Vol. 1; Amsterdam, etc.: Elsevier, 2012. pp. 603-651

[96] Kabanov P, Baranova D. Cyclothems and stratigraphy of the Upper Moscovian-Basal Kasimovian succession of central and northern European Russia. In: Wong T, editor. Proc. 15 Int. Cong. Carbonif. Perm. Amsterdam: Royal Netherlands Academy of Arts and Sciences; 2007. pp. 147-160

[97] Kabanov P, Alekseev AS. Progress in cyclothem/sequence stratigraphy of type Lower Moscovian succession of Moscow Basin, Russia. Newsletter on Carboniferous Stratigraphy. 2011;**29**:42-50

[98] Baranova DV, Kabanov PB, Alekseev AS. Fusulinids, facies, and biofacies of the upper Moscovian (Carboniferous) of southern Moscow Basin and Oka-Tsna Swell. Paleontological Journal. 2014;**48**:701-849

[99] Kabanov PB, Alekseev AS, Gibshman NB, Gabdullin RR, Bershov AV. The upper Viséan–Serpukhovian in the type area for the Serpukhovian Stage (Moscow Basin, Russia): Part 1. Sequences, disconformities, and biostratigraphic summary. Geological Journal. 2014;**51**:163-194

[100] Alekseeva TV, Kabanov PB, Zolotareva BN, Alekseev AO, Alekseeva VA. Humic substances of the late Carboniferous palygorskitic paleosol from the Southern Moscow region, Russia. Doklady Biological Sciences. 2009;**425**:128-132

[101] Platt NH, Wright VP. Palustrine carbonates and the Florida Everglades: Towards an exposure index for the fresh-water environment?. Journal of Sedimentary Petrology. 1992;**62**:1058-1071

[102] Alekseeva TV, Alekseev AO, Gubin SV. Paleosol complex in the uppermost Mikhailovian Horizon (Viséan, Lower Carboniferous) in the southern flank of the Moscow Syneclise. Paleontological Journal. 2016;**50**:319-335

[103] Alekseeva TV, Alekseev AO, Gubin SV, Kabanov PB, Alekseeva VO. Palaeoenvironments of the Middle–Late Mississippian Moscow Basin (Russia) from multiproxy study of palaeosols and palaeokarsts. Palaeogeography, Palaeoclimatology, Palaeoecology. 2016;**450**:1-16

[104] Morrow DW. Devonian of the Northern Canadian Mainland Sedimentary Basin (a contribution to the Geological Atlas of the northern Canadian Mainland Sedimentary Basin). Geological Survey of Canada, Open File. 2012;**6997**:88

[105] Kabanov PB. Landry Formation of Kugaluk N-02 well (Devonian, northern mainland NWT): Insight into formation's boundaries, lithofacies, and stratal stacking patterns. Bulletin of Canadian Petroleum Geology. 2014;**62**:120-139

[106] Kabanov P. Geological and geochemical data from Mackenzie Region. Part I. Devonian cored sections and new geochemical, δ13C– δ18C, and pyrolysis data. Geological Survey of Canada Open File. 2015;**7840**:94

[107] Gal LP, Pyle LJ, Hadlari T, Allen TL. Chapter 6 – Lower to Upper Devonian strata, Arnica – Landry Play, and Kee Scarp Play. In: Pyle LJ, Jones AL, editors. Regional Geoscience Studies and Petroleum Potential, Peel Plateau and Plain, Northwest Territories and Yukon. Project Volume. NWT Open File Report 2009-02 and YGS Open File 2009-25; Yellowknife, NT: NTGO, 2009. pp. 187-289

[108] Gal LP, Pyle LJ. Chapter 5 – Upper Silurian – Lower Devonian strata (Delorme Group). In: Pyle LJ, Jones AL, editors. Regional Geoscience Studies and Petroleum Potential, Peel Plateau and Plain, Northwest Territories and Yukon. Project Volume. NWT Open File Report 2009-02 and YGS Open File 2009-25; Yellowknife, NT: NTGO, 2009. pp. 161-186

[109] Kabanov P, Gouwy SA, Chan WC. Geological and geochemical data from Mackenzie Corridor. Part VI: Descriptions and SGR logs of Devonian outcrop sections, Mackenzie Mountains, Northwest Territories, NTS 106G and 106H. Geological Survey of Canada, Open File. 2016;**8173**:94

[110] Vopni LK, Lerbekmo JF. Sedimentology and ecology of the Horn Plateau Formation: A Middle Devonian coral reef, Northwest Territories, Canada. Geologische Rundschau. 1972;**61**:626-646

[111] Corlett H, Jones B. The influence of paleogeography in epicontinental seas: A case study based on Middle Devonian strata from the MacKenzie Basin, Northwest Territories, Canada. Sedimentary Geology. 2011;**239**:199-216

Chapter 7

Integrated Stratigraphy of the Cenozoic Andean Foreland Basin (Northern Argentina)

Claudia Inés Galli, Ricardo Narciso Alonso and
Lidia Beatriz Coira

Additional information is available at the end of the chapter

http://dx.doi.org/10.5772/intechopen.69985

Abstract

The stratigraphic and sedimentologic characteristics of Cenozoic deposits in north-western Argentina represent important tectono-sedimentary constraints on the evolution of the Andean foreland basin in this region. This nonmarine unit unconformably rests on the top of the postrift deposits of the middle Eocene Lumbrera Formation (Santa Bárbara Subgroup, Salta Group) or on older deposits. Eocene-Pliocene paleoenvironmental changes were the direct result of changes in the tectonic setting and accommodation space. This study describes the results of an integrated analysis of the middle-upper Eocene to Plio-Pleistocene deposits filling the basins of the Cordillera Oriental. Fluvial deposits associated with different topographic slopes characterize the basins that formed in the Central Andes of north-western Argentina due to Cenozoic tectonic convergence. The formation of these basins led to the development of continental sedimentary environments, including an ephemeral fluvial system with aeolian dune fields; a sandy braided fluvial system; a playa lake; a sinuous gravelly sandy fluvial system with lagoons; and lagoons and marshes. These basins, which were probably connected during the first stage of their development, are characterized by different subsidence histories, sedimentary paleoenvironmental evolution patterns, topographic slopes, provenances, and paleocurrent directions, resulting in different tectono-sedimentary histories.

Keywords: Andean foreland basin, Payogastilla Group, Orán Group, provenance, stratigraphy, sequence stratigraphic, magnetostratigraphy

1. Introduction

The two-dimensional elastic model of the evolution of foreland basins, which proposes that the thrust load and the sedimentary load produce a wide deflection in the lithosphere [1, 2],

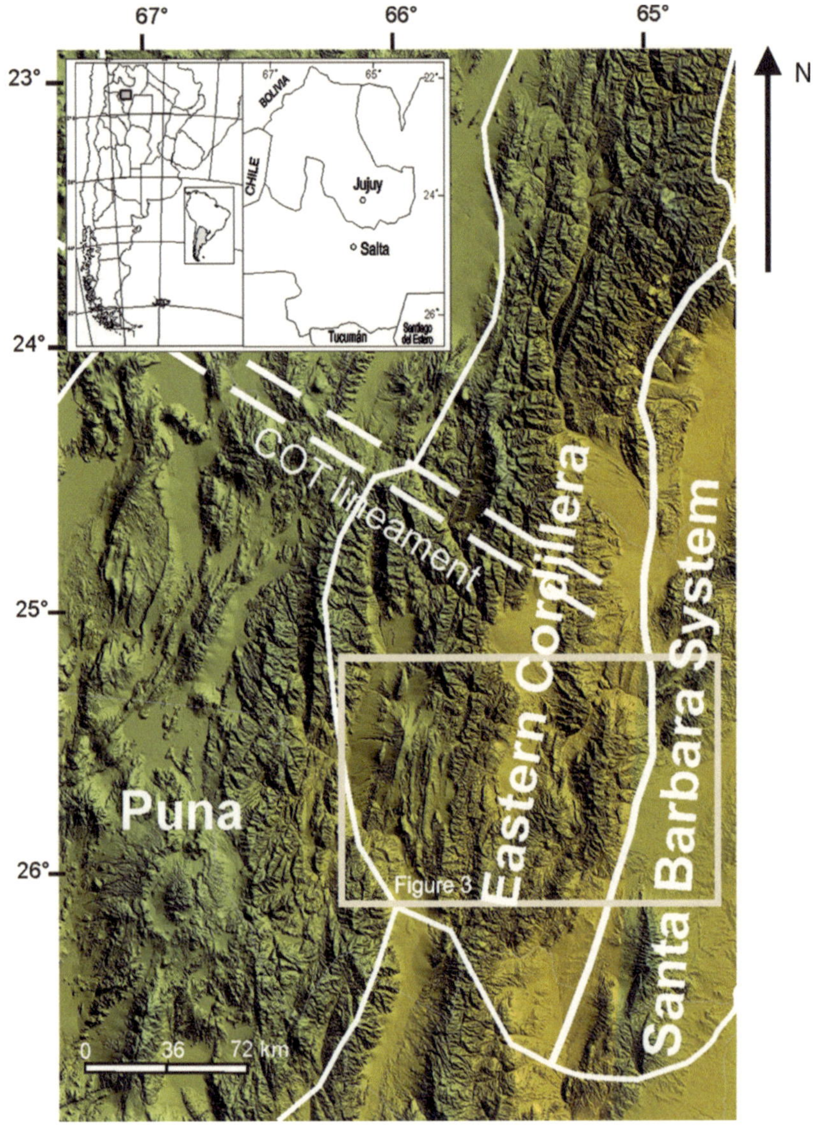

Figure 1. Satellite image of regional location of Cenozoic foreland basin.

has been widely applied to foreland basins [3–9]. These models show that the tectonic activity and the evolution of the fold and thrust belt are the drivers of subsidence in the associated foreland basins [10]. Sediment redistribution, autocyclic sedimentary processes, and eustatic baseline changes are important factors affecting the basin characteristics [2].

The main elements of "foreland basin systems" result from the accommodation created by the bending of the crust in response to the topographic load of the fold and thrust belt. This widely applied model includes four depozones: wedge-top, foredeep, forebulge, and back bulge [6].

Another model in the northern Argentina, the "broken foreland basin," describes basins that form in retroarc zones that are largely influenced by basement structures. In these basins, the accommodation develops mainly along reactivated and inverted structures, thereby giving rise to relatively restricted basins with variable and laterally limited connections [10].

The clastic deposits of the Middle Eocene-Pliocene are excellent examples of the Cenozoic foreland basin, which is associated with Andean orogeny [11, 12] that evolved into intermontane basins (**Figures 1** and **2**). In north-western Argentina, the Cenozoic sediments reflect the passage of a "rift" basin during the Eocene. These deposits are associated with the Salta Group and the Andean foreland basin, which comprises extensive basins in which the Payogastilla

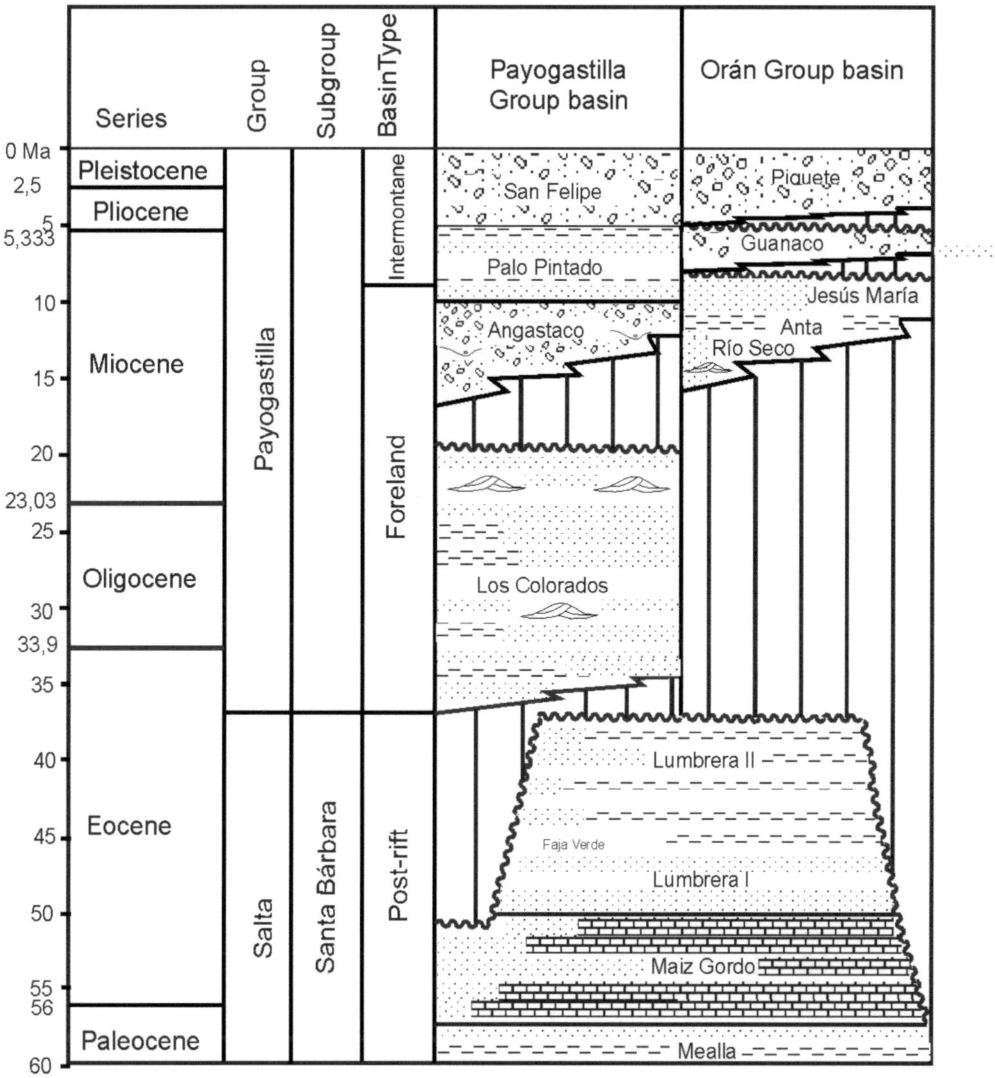

Figure 2. Stratigraphic chart of the units from Calchaquí basin and Orán basin.

Group (Valle Calchaquí) and the Orán Group (Lerma Valley, Sianca Valley, Santa Bárbara System, and Sierra de Zapla) accumulated (**Figures 1** and **2**).

The structural evolution of the Andean foreland basin was mainly controlled by the inversion of the extensional basins of the Cretaceous rift of the Salta Group, which overlaps with the general migration of the deformation toward the foreland. Some authors describe this as "broken foreland basin," a view with widespread consensus today [13, 14], while others refer to it as a "foreland basin system" [6, 7].

The Cenozoic Andean foreland basin provides an excellent opportunity to define the relationships between tectonism and sedimentation because its geologic history is closely linked to the tectonic activity over the evolution of the river system in the basin. In this study, the paleoenvironmental characteristics, the types of contacts between units, the provenance of the deposits, and the geochronologic and paleomagnetic ages of the stratigraphic units in each basin are presented. The integration of these data has improved our understanding of the basin evolution during the Andean orogeny.

2. The pre-Andean basement

In Cordillera Oriental, the upper Neoproterozoic La Paya Formation basement unit contains low-grade metamorphosed sandstones and mudstones [15, 16] that grade southward into schists, gneisses, and migmatites [17, 18] in the Sierra de Quilmes and Cumbres Calchaquíes (**Figures 2** and **3**).

Marine quartzites of the Meson Group are arranged in angular unconformity on top of the previous deposits [19] (middle to upper Cambrian).

The marine deposits of the Silurian-Devonian basin are represented by deposits of an extensive marine platform environment whose greater thicknesses are developed east of the Cordillera Oriental [20].

The sedimentary succession that overlaps the Neoproterozoic to lower Paleozoic basement corresponds to the Cretaceous-Paleogene strata of the Salta Group [21] and the Paleogene-Neogene strata of the Payogastilla Group and Orán Group.

The Salta Group, in Cordillera Oriental and Santa Bárbara System, is present in three subbasins: Metán, Alemanía, and Pucará-Brealito (**Figure 3**). The Salta Group deposits are divided into the following three subgroups (from base to top): Pirgua [22], Balbuena, and Santa Bárbara [23]. The Pirgua Subgroup is composed of sandstones, conglomerates, and siltstones at almost all localities and represents the syn-rift fill. The Balbuena Subgroup, which accumulated during the Maastrichtian to Early Paleocene, represents the early postrift stage and is composed of white sandstones (Lecho Formation) and gray to yellow limestones in the upper part (Yacoraite Formation). The Santa Bárbara Subgroup consists of the Mealla, Maíz Gordo, and Lumbrera formations [23] and is dominated by fine-grained red sandstone, siltstone, and green mudstone.

The Lumbrera Formation [23], which represents the uppermost part of the Salta Group (**Figure 2**) is composed of claystones and siltstones and is always reddish-brown to red. In the

http://dx.doi.org/10.5772/intechopen.69985

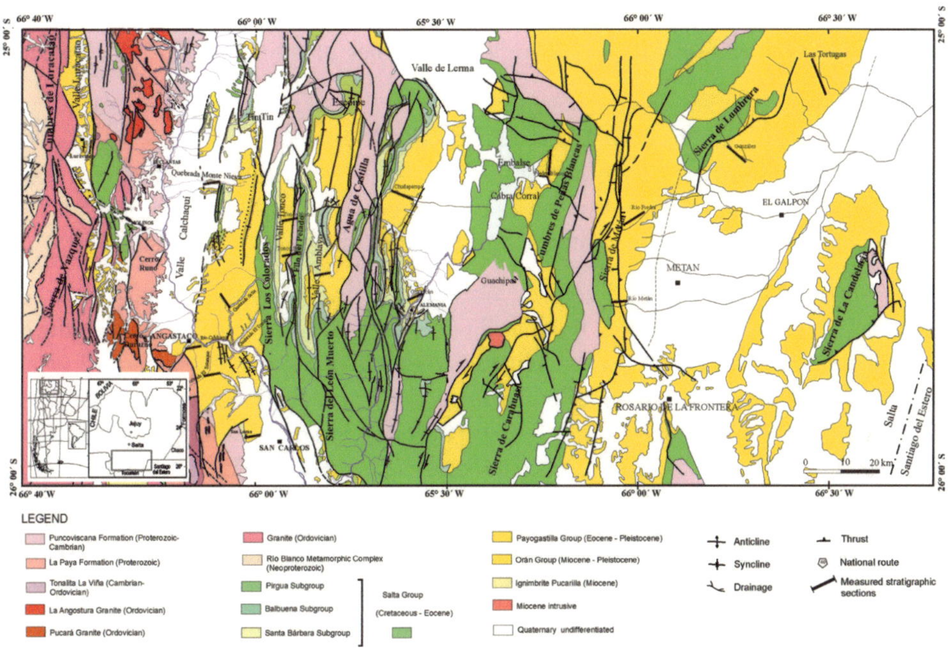

Figure 3. Geologic map of the region in southern Salta province, northwest Argentina. Adapted from [68, 69].

Lumbrera Formation, three units have been identified based on the contrasting characteristics of the facies groups: the lower Lumbrera Member, the *Faja Verde*, and the upper Lumbrera Member [24]. The lower part is fossil-rich [25–27] and was dated to the lower to middle Eocene based on vertebrate associations [28]. The top of the Faja Verde is an omission surface that marks the beginning of the sedimentary foreland basin for some authors [4, 29–31].

A clear paraconformity is located between the Lumbrera II Formation and the Los Colorados Formation. In parts of the foreland basin, a paraconformity between Lumbrera II Formation and Río Seco Formation corresponds to a hiatus from lower Eocene to middle Miocene.

The Orán Group [32] includes the Paleogene and Neogene deposits, which, in the Salta and Jujuy provinces and surrounding areas, overlie on the Salta Group (Cretaceous-Eocene, **Figure 2**). During the upper Neogene, several tectonic events related to the uplift of the Andes occurred and caused variations in the basins located to the east, consequently influencing the characteristics of the sediments, which can be divided into two sequences: the lower, Metán Subgroup, and the upper, Jujuy Subgroup (**Figure 2**).

3. Calchaquí Basin

3.1. Los Colorados Formation

The middle to upper Eocene deposits of the Payogastilla Group, including the Los Colorados Formation, represents the initial stage in the evolution of the Andean foreland basin of north-western Argentina (**Figure 2**). The area of Tin Tin, Tonco, and Calchaquí valleys feature outcrops with well-documented, complete profiles of the Los Colorados Formation (**Figure 4**).

The first episode of filling of the basin started with the deposition of the Los Colorados Formation in the middle to upper Eocene [33, 34]. A clear second-order subaerial unconformity Type 2 [35, 36] is located between the Lumbrera Formation and Los Colorados Formation. This unconformity is also found in the Luracatao and Pucará valleys [13, 14, 31, 34].

Based on the fossil record, the initial development of the foreland basin, at least in the Luracatao Valley, occurred during the middle Eocene [34]. Sedimentary filling of the basin began with the deposition of the Los Colorados Formation and was characterized by sheetflood ephemeral fluvial deposits formed by unconfined and confined channels within dune fields in an arid region. In some parts of the basin, aeolian deposits interfinger with ephemeral fluvial systems, such as those in the Tonco Valley (Sequences I and III).

Detrital zircons from the town of Angastaco have been dated to 37.6 ± 1.2 Ma [37] and the apatites from Monte Nieva have been dated to 28.7 ± 1.9 Ma [6]. This deposit is ~300 m thick and contains basal sandstone and siltstone facies of an ephemeral fluvial system and is associated with aeolian deposits [38]. A tuff horizon intercalated in the aeolian deposits in the Tin Tin section has provided an age of 21.0 ± 0.8 Ma (U-Pb) [39].

3.1.1. Facies, depositional architecture, and sequence stratigraphy of the Los Colorados Formation

The Los Colorados Formation is an unconformity-bounded depositional sequence that corresponds to a major stratigraphic cycle in the evolution of the foreland basin in the Calchaquí

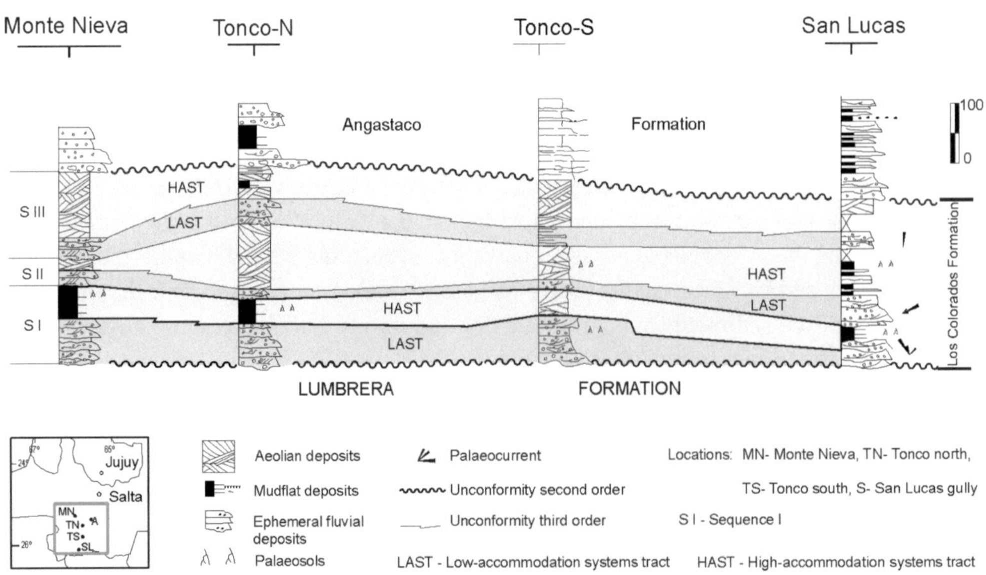

Figure 4. Correlation of Los Colorados Formation based on the sequence stratigraphy analysis.

Valley. The entire Payogastilla Group corresponds to a first-order sequence [35, 36], i.e., it is associated with one distinct tectonic setting. Hence, according to the hierarchy based on the magnitude of base-level changes that resulted in the formation of the sequence, the Los Colorados Formation can be assigned to a second-order level of stratigraphic cyclicity. In this context, the three depositional sequences that form its stratigraphic subdivisions can be considered third-order sequences. The correlations reveal that the deposits are separated at the base and top by second-order subaerial unconformities [35, 36].

The lower boundary of the Los Colorados Formation is marked by an increase in the grain size, by a paleoenvironmental change from the mud flat deposits of the Lumbrera Formation to ephemeral fluvial systems with conglomerates and by a marked change in the sedimentary provenance. This unconformity is very clear in the northern part of Amblayo Valley. The controversial upper boundary is an erosional unconformity (Tonco Valley) and represents a change in depositional paleoenvironment from distal sandy ephemeral fluvial system and clay playa deposits or aeolian accumulations to a braided fluvial system (**Figures 4** and **5**).

The Los Colorados Formation deposits were identified as facies of an ephemeral fluvial system with flashy discharge, calcic paleosols, and dune fields, which are characteristic of arid regions.

Sequence stratigraphic concepts are applicable, with modifications, to the successions that are entirely nonmarine in origin, even where there are no marine surfaces with which to correlate them, such as in the Payogatilla Group basin. In a fully nonmarine environment, fluvial accommodation is created and destroyed by the following: (a) differential tectonic movement between basin and source areas, which can modify the amount of sediment supply and the gradient of the landscape profile and (b) cycles of climate change, which can alter the balance between fluvial discharge and sediment load [40].

3.2. Angastaco Formation

The thickness of the Angastaco Formation varies considerably between the sections in which it is fully exposed (e.g., from 4450 m at the Calchaquí River to 1500 m in the Tonco Valley, **Figure 6**). The depocenter of the basin between ~13.7 and 10 Ma was located in the area of Angastaco.

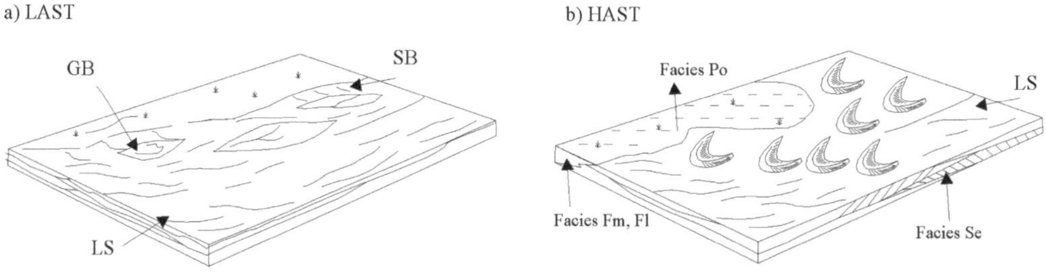

Figure 5. Palaeoenvironment with architectural elements illustrated using schematic diagrams of Los Colorados Formation. a) LAST (e g. Sequence II), proximal ephemeral confined (SB element) in the base, b) unconfined ephemeral – mud flat (LS element) associated with aeolian deposits (HAST).

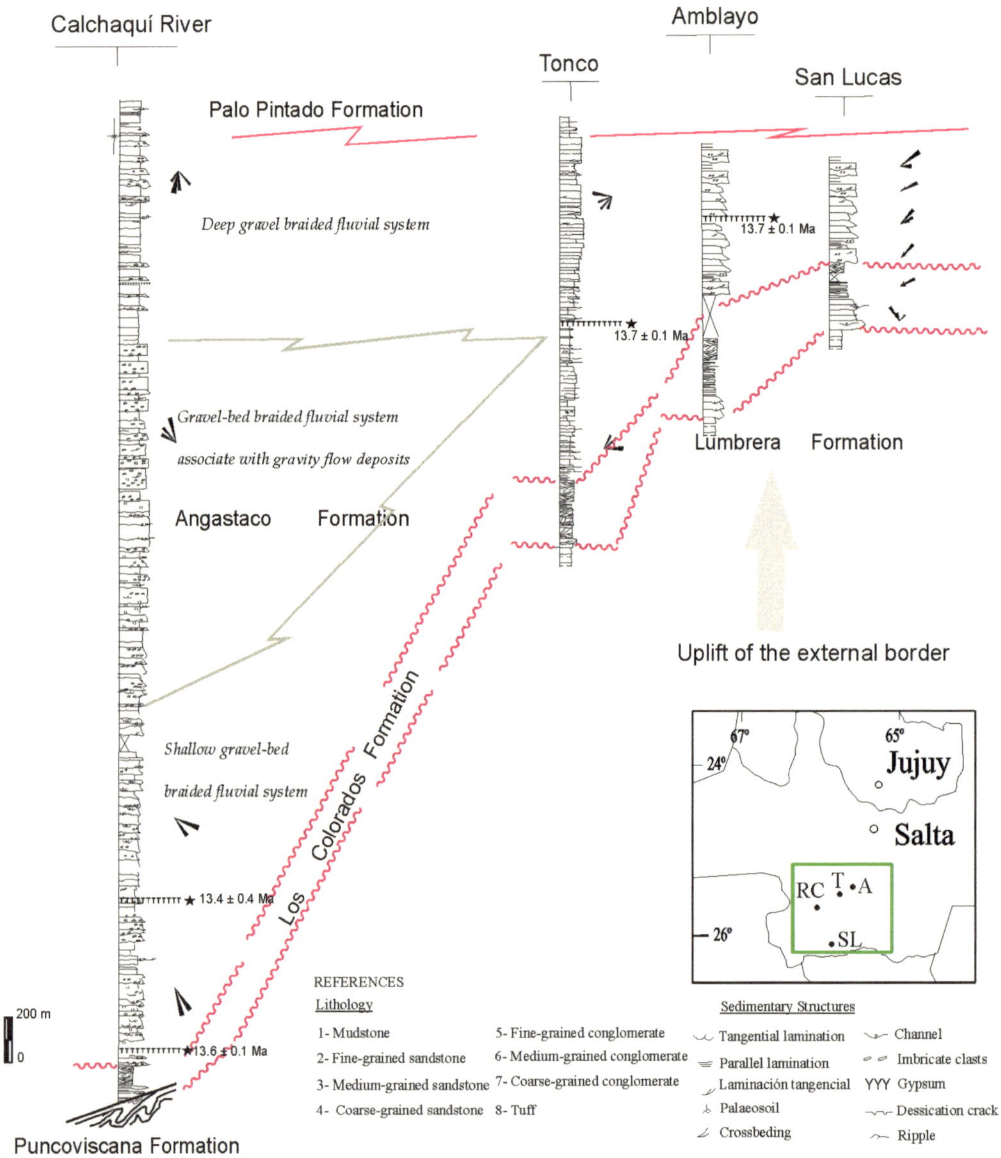

Figure 6. Stratigraphic correlation of the Los Colorados and Angastaco formations (base of Pavogastilla Group).

The structures are located on the western edge of the basin and are similar to those that created local accommodation space in other broken foreland settings [10, 13, 14, 41, 42].

3.2.1. Facies and depositional architecture of the Angastaco Formation

Several fluvial systems have been recognized based on the lithofacies and stratigraphic architectural. The lithofacies were characterized based on the deposits' properties (**Table 1**) and on the stratigraphic analysis (**Figure 7**, **Table 2**).

http://dx.doi.org/10.5772/intechopen.69985

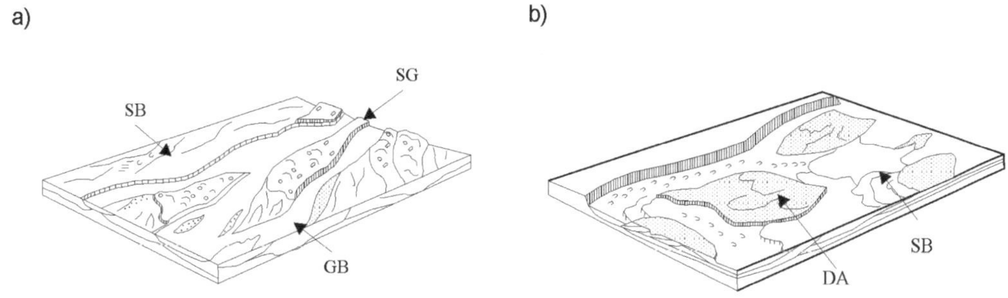

Figure 7. Description of the palaeoenvironment with architectural elements illustrated using schematic diagrams of Angastaco Formation. a) Gravel-bed braided river system associated with gravity flow deposits, Angastaco Formation of the middle part of the deposits with its architectural elements, b) Deep gravel-bed braided river at the top Angastaco Formation, with its architectural elements.

Code	Lithofacies	Interpretation
Gmm	Massive conglomerate, poorly sorted matrix-supported clasts, sandy matrix. Strata with abrupt lateral terminations.	Debris flow (high strength). Flows passively occupy preexisting alluvial topography and mold to the preexisting channel.
Gmg	Massive conglomerate, matrix-supported clasts with normal grading.	Pseudoplastic debris flow (low strength).
Gcm	Clast-supported, massive conglomerate, with poorly sorted, very angular clasts.	Pseudoplastic debris flow (inertial bedload).
Gh	Clast-supported conglomerate, crudely bedded, sandy matrix. Imbrication.	Longitudinal bars, lag deposits.
Gi	Conglomerate with well-sorted and rounded clasts, sparse matrix, imbrication.	Longitudinal bars.
Gt	Matrix-supported conglomerate with trough cross-bedding.	Minor channel fills.
Se	Sandstone with cosets of grouped trough cross-beds (10–20 m thick), very well-sorted, rounded grains.	Aeolian dune deposits.
St	Very coarse-grained sandstone, sets of trough cross-beds. Wedge-shaped strata with residual lag.	Linguoid (3D) dunes.
Sp	Very coarse-grained sandstone, sets of planar cross-beds, and lag deposits. Beds with erosional bases.	Transverse and linguoid bedforms (2D dunes).
Sl	Fine- to coarse-grained sandstone, planar lamination, lag deposits. Tabular beds.	High flow conditions. Flash flood.
Sm	Fine- to coarse-grained sandstone, poorly sorted, massive, with clastic wedges and beds with erosional bases.	Upper flow regime and poorly sorted deposits.

Code	Lithofacies	Interpretation
Fl	Very fine-grained sandstone, siltstone, and mudstone, fine lamination, desiccation cracks, roots, bioturbation.	Overbank, abandoned channel, or waning flood deposits.
Fm	Massive siltstone and mudstone, very thin beds.	Overbank or abandoned channel deposits.
Fo	Siltstone and mudstone, very small ripples and very thin laminations.	Swamp and lacustrine in the floodplain.
Po	Very fine-grained sandstone, siltstone and mudstone, massive with calcified rhizoliths penetrating down into sandstone of aeolian dune origin.	Paleosols. The rhizoliths emanate from the bases of damp and wet interdune units.

Table 1. Major lithofacies identified in the Payogastilla Group and Oran Group (modified from Ref. [47]).

Formation	Code	Architectural elements	Principal lithofacies
San	GB	Gravel bars	Gmg - Gi - Gt - Sm
Felipe	SB	Sandy bedforms	St - Sm - Fm
	GB	Gravel bars and bedforms	Gt - Gm
Palo	LA	Lateral-accretion macroform	Gt - Sm
	CS	Crevasse splay - Channel	Sp - Sl - Sm
Pintado	FF(CH)	Abandoned channel fills	Fl - Fm
	FF	Foodplain deposits	Fl - Fm - Fo - Po
	SB	Sandy bedforms	Sl - Sm - Fl
	GB	Gravel bars and bedforms	Gh - Sm
	DA	Downstream – accretion macroform	Gh - Gi - Gm - Sl - Sm - St
	SB	Sandy bedforms	Sm - St - Sp - Sm
Angastaco	GB	Gravel bars and bedforms	Gh - Gcm
	SG	Sediment gravity flows	Gmg
			Gh - St
	GB	Gravel bars and bedforms	Gh - Gi - Sm
			Gh - Gt - Gmg
			Gh - Gt - Sl
	SB	Sandy bedforms	Sl - St - Sp - Fl - Fm
Los Colorados	LS	Laminated sand sheets associated with aeolian deposits	Sl - Fm - Fl - Se
	SB	Sandy bedforms	Gcm - Sm - St - Sp
	GB	Gravel bars and bedforms	Gm - Gt

Table 2. Codes of the major architectural elements defined for the Payogastilla Group with their characteristic lithofacies.

http://dx.doi.org/10.5772/intechopen.69985

The Angastaco Formation conglomerates in the western part of the basin contain substantial amounts of plutonic rocks from the Oire Eruptive Belt. In the eastern part of the study area (Tonco profile), however, slates, phyllites, and schists from the Puncoviscana Formation are present. There are fewer paleovolcanic clasts from the Eruptive Belt of the eastern Puna (Calchaquí River) and neovolcanics in the San Lucas River and the Tonco south area, which are associated with paleocurrent directions from the north-west and west. In the San Lucas and the Tonco south River area, a small but significant component of red sandstone clasts from the Formation and gray sandstones from the Maíz Gordo Formation represent the Salta Group. These data suggest the tectonic uplift of the Sierra León Muerto in the eastern study area (**Figures 3** and **6**) [43].

In the upper section, the paleoenvironment changes to more erosional rivers with deep channels. The paleocurrents are from the north-west and are associated with neovolcanic clasts from the volcanic-sedimentary deposits of the Almagro-El Toro basin, which has a depositional and eruptive age between 14.3 and 6.4 Ma and synorogenic deposits dated at ~11 Ma [44].

3.3. Palo Pintado Formation

The Palo Pintado Formation is ~800 m thick and contains a tuff level that has been dated to 10.29 ± 0.11 Ma (K/Ar) [45]. Near the top is another pyroclastic level that has been dated to 5.27 ± 0.28 Ma ($^{206}Pb/^{238}U$) [46] and 5.98 ± 0.32 Ma [47] (**Figure 7**). The unit comprises thickening- and coarsening-upward cycles, including matrix-supported conglomerates, fine- to medium-grained sandstones, and fine-grained sublithic sandstones ending in green, brown, and gray siltstones levels (**Figure 8a**).

These deposits have been interpreted as wandering sand-gravel fluvial systems with small lakes [48]. The geometry and the fluvial architectural characteristics are a direct consequence of allogenic controls, such as tectonic activity, under constant climatic conditions.

During the upper Miocene, the uplift of the basin caused an increase in the sedimentary accommodation/deposition (A/D) rate and was also associated with a change in the petrologic composition of the deposits [48]. The resulting orographic barriers produced a warmer and wetter climate [49].

3.3.1. Facies and depositional architecture of the Palo Pintado Formation

The fluvial architectural characteristics and associated lithofacies in the Palo Pintado Formation define a fluvial system with intrachannel and overbank deposits [43]. The intrachannel deposits include gravel bars and bedform deposits (GB) and sandy bedforms comprising transverse bars and sand waves formed by vertical accretion and downstream flow (SB) (**Table 2**, **Figure 8a**). In contrast, the overbank deposits are represented by three types of features: (a) lateral accretion macroforms, which are characterized by large-scale, gently dipping second-order bounding surfaces that correspond to successive increments of lateral growth, with erosional bases and gradational tops; (b) small crevasse splay channels resulting from erosion at

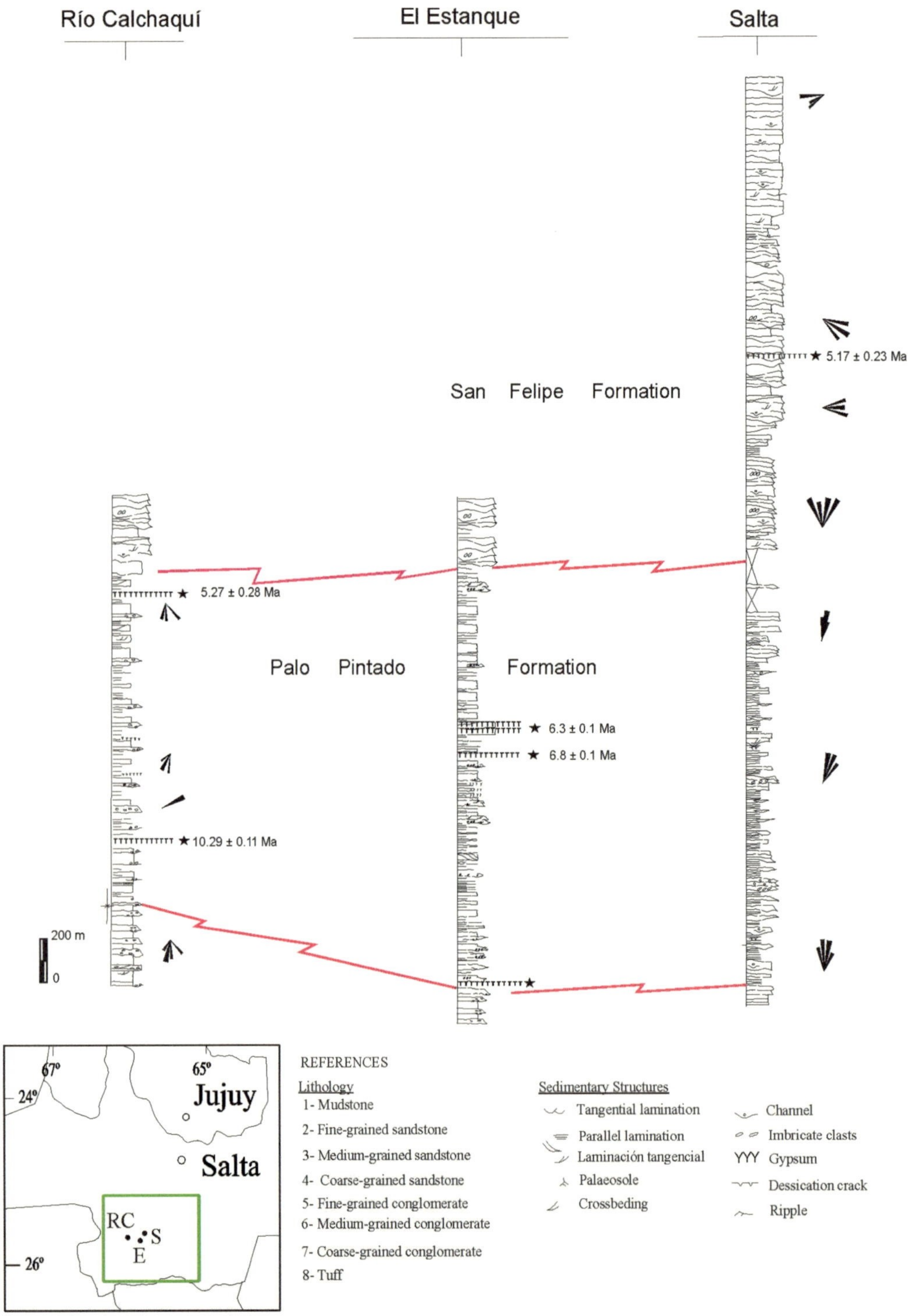

Figure 8. Stratigraphic correlation of the deposits of Palo Pintado and San Felipe Formations (top of Payogastilla Group).

the borders of the main channel during flood events, which correspond to crevasse channels (CS); and (c) the development of large floodplain deposits (FF) (**Figure 8a**, **Table 2**).

X-ray diffraction data from floodplain clay minerals revealed the presence of illite, montmorillonite, magnesium-rich smectite, and kaolinite generated by hydrolysis under a warm and humid climate [50].

The occurrence of *Caiman* cf. *latirostris* also supports the hypothesis that the climatic conditions in Valle Calchaquí during the upper Miocene were comparatively wetter than those inferred for contemporaneous units deposited to the east (Guanaco Formation, Orán Group) [51].

The paleomagnetic analysis reflects the increase in the sedimentation rate from 0.41 mm/year at the base, to 0.11 mm/year at the middle, to 0.66 mm/year in the top of the deposits, which is associated with a higher percentage of Salta Group clasts. Paleocurrent directions from the south and the south-east indicate the tectonic reactivation of the deposition area from the Sierra León Muerto (and its continuation to the north as the Sierra Los Colorados). The exhumation was registered before in the conglomerates of the Angastaco Formation [43, 52].

In the Quebrada Salta, quartzite clasts with *Skolithos* from the Mesón Group (upper Cambrian) and paleocurrents from the north and north-east suggest a provenance from Quebrada El Toro, where the Mesón Group (upper Cambrian) is well exposed.

3.4. San Felipe Formation

The deposits of the San Felipe Formation at the top of the Payogastilla Group are more than 600 m thick in the south-eastern Calchaquí Valley and are affected by numerous faults and folds. The transition between the Palo Pintado Formation and the San Felipe Formation is sharp and unconformable. The outcrops of San Felipe Formation, present less areal distribution than the previous and are restricted to the south-eastern sector of the Calchaquí basin (**Figure 8**).

3.4.1. Facies and depositional architecture of the San Felipe Formation

The San Felipe Formation is characterized by conglomerates deposited in low-sinuosity channels.

The well-sorted conglomerates lack of matrix, contain rounded clasts overlapping in thick tabular strata, are 2 to >7 m in thickness, and are found in longitudinal bar deposits (**Table 1**, **Figure 9b**). The unit also contains poorly sorted conglomerates, supported clasts, pseudoplastic debris flow, and massive coarse-grained wackes resulting from rapid accumulation and poorly sorted deposition. The origin of the conglomerates in the San Felipe Formation was analyzed in the Quebrada Salta, where there are also elements of the Puncoviscana Formation and the Oire Eruptive Complex. In addition, limestone clasts from the Yacoraite Formation (Salta Group) have been found and clasts from Pirgua Subgroup (Salta Group) in association with paleocurrent directions from the west and south-west. The San Felipe Formation deposits have been interpreted as braided alluvial fans associated with shallow gravelly braided fluvial system (**Figure 9b**, **Table 2**) [38].

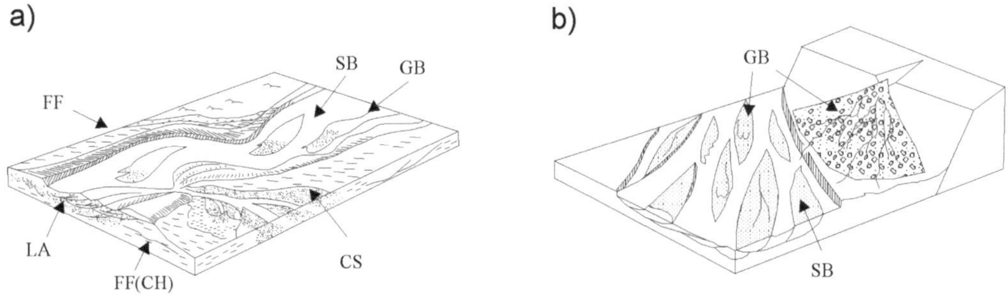

Figure 9. Description of the palaeoenvironment with architectural elements illustrated using schematic diagrams. a) meandering fluvial sand-gravel system with small lakes of the Palo Pintado Formation with its architectural elements, and b) braided fluvial fan and river system of San Felipe Formation with its architectural elements.

4. Orán Group

4.1. Metán Subgroup (Río Seco, Anta, and Jesús María formations)

The basal contact of the Metán Subgroup deposits corresponds to a regional unconformity and is generally associated with the Lumbrera Formation (Santa Bárbara Subgroup, Salta Group). In the eastern part of the basin (Umbral de Los Gallos [53]), this subgroup lies on different units of the postrift deposits of the Salta Group, such as the Lumbrera, Maíz Gordo, Mealla, and Yacoraite formations (**Figure 7**). Although the distribution of the Metán Subgroup is broader than that of the Salta Group, its depocenters are generally the same as those existing during the accumulation of the Salta Group [54].

In this basin, a zircon fission-track sample from an intercalated tuff at Alemanía yielded an age of 14.5 ± 1.4 Ma and hornblende crystals from a tuff collected in the lower part of the Anta Formation at Río Piedras produced a $^{40}Ar/^{39}Ar$ date of 13.95 ± 0.72 Ma. The age of the lowest exposed Anta Formation beds is 15.2 Ma at Río Piedras and 17.3 Ma at Río Metán based on magnetic stratigraphy [55, 56]. Other authors have placed the contact with the overlying Guanaco Formation at 12.3 Ma (Arroyo Piedra Blanca), 13.5 Ma (Río Metán), 13.1 Ma (Río Piedras), and 9.7 Ma (Arroyo González) [56, 57] (**Figure 7**). The paleomagnetic ages obtained from the contacts at the base and top of the Metán Subgroup vary in different parts of the basin, from between 17.3 and 12 Ma to between 15 and ~9 Ma (**Figure 7**). These data reveal that an elongated initial depocenter developed parallel to the uplift zone at ~17 Ma, migrated toward the eastern edge of the recent basin at ~15 Ma and continued to fill with sediment until ~12 Ma, when new basin structuring and the erosion of the Metán Subgroup deposits began.

4.1.1. Facies and depositional architecture of the Metán Subgroup

The Río Seco, Anta, and Jesús María formations present interfingering stratigraphic relationships (**Figure 11**). The deposits of the Metán Subgroup (Río Seco and Jesús María formations) are characterized by a succession of lithofacies from fine-grained to very coarse-grained sandstone, often with pelitic clasts (**Table 3**) and have been interpreted as the product of ephemeral flows that deposited sand sheet under high-flow regime conditions (**Figures 10** and **11**, **Table 3**).

The deposits of the Río Seco and Jesús María formations have been interpreted as accumulates in a paleoenvironment of a "sandy ephemeral fluvial system associated with dune fields" under arid climatic (**Figure 11**) [54, 56].

The Anta Formation is primarily composed of brown, green, and yellow mudstone and medium- to fine-grained sandstone. Gypsum layers, gypsum nodules, and pyroclastic layers are common throughout the basin. The oolitic limestones with foraminifera are found in the south-eastern sector of the basin. These limestones have been assigned to the Paraná marine ingression with 14.9 Ma in age, based on paleomagnetic data collected at the Piedras river (Miliolidos, **Figure 10**) [54, 56].

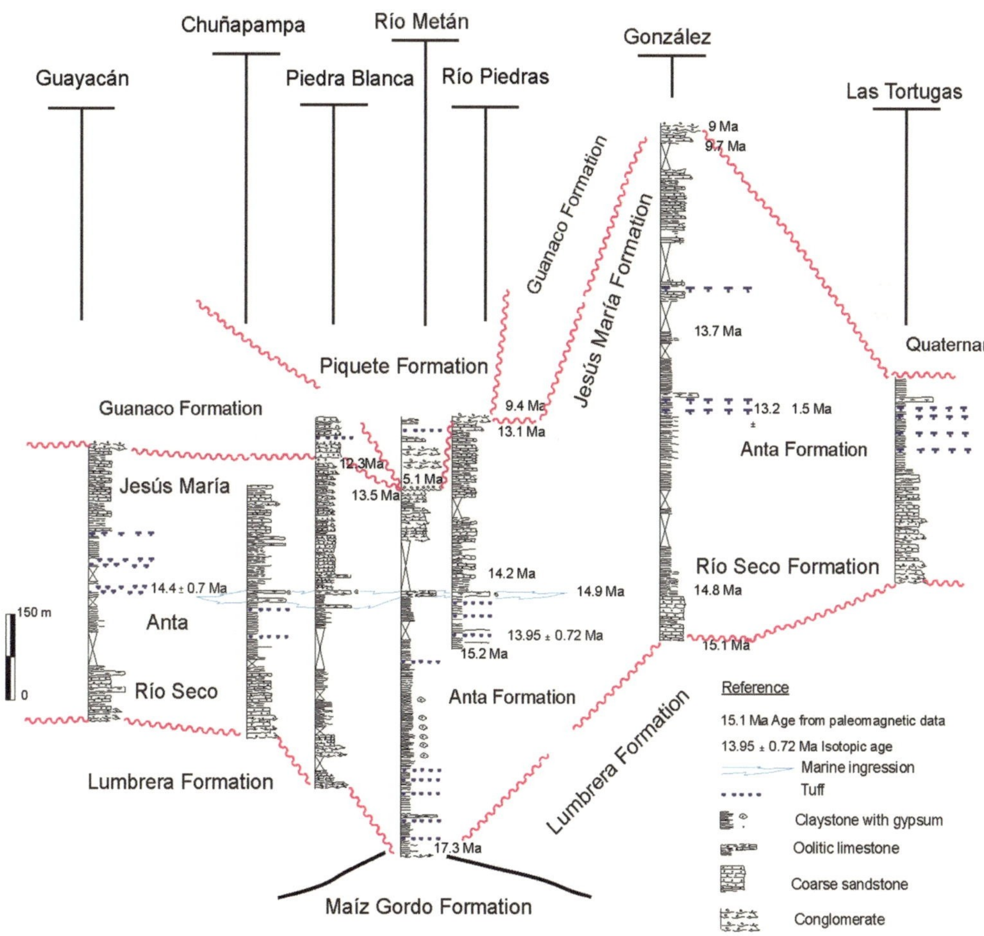

Figure 10. Stratigraphic correlation of the deposits of Metán Subgroup (base of Orán Group).

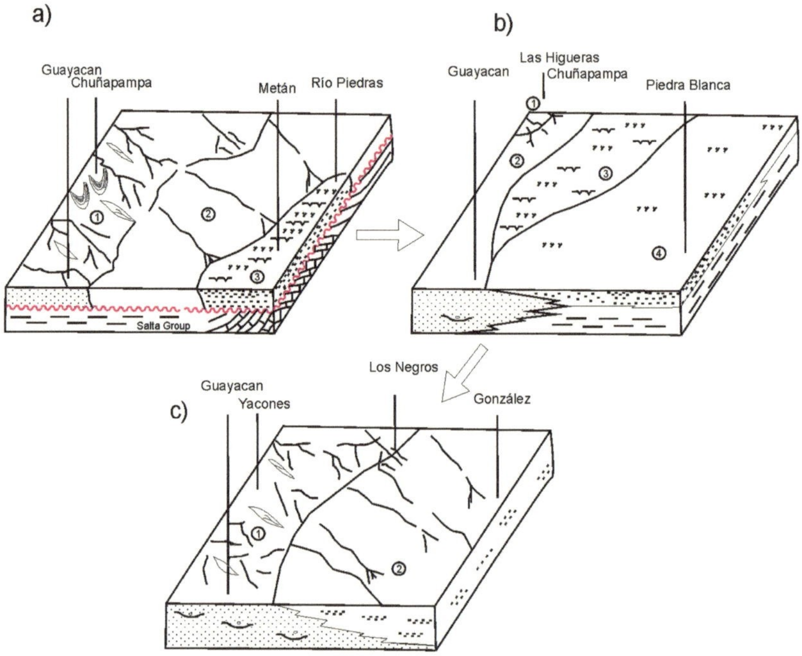

Figure 11. Description of the palaeoenvironment with architectural elements illustrated using schematic diagrams of Metán Subgroup a) 1 - proximal ephemeral sandy fluvial system associated with wind deposits, 2- distal ephemeral sandy fluvial system, 3- playa lake; b) 1- alluvial fan deposits, 2- sandy plain, 3- dry mud flat, 4- ephemeral saline lake; c) 1- proximal ephemeral fluvial system and, 2- distal ephemeral fluvial system.

Formation	Code	Architectural elements	Principal lithofacies
Piquete	SG	Sandy bedforms	Gcm - Sm - St - Sp
	GB	Gravel bars	Gmg - Gi - Gt - Sm
Guanaco	GB	Gravel bars	Gmg - Gi - Gt - Sm
	SB	Sandy bedforms	St - Sm - Fm
	CH	Channel	
Jesús	FF	Foodplain deposits	Fl - Fm - Fo - Po
María	LS	Laminated sand sheets	Sl - Fm - Fl - Se
	SB	Sandy bedforms	Gcm - Sm - St - Sp
	FL	Saline lake	Fl - Fm - Co
Anta	FF	Mud Flat	Fl - Fg - Fgr
	SB	Sand Flat	Sm - St - Sp - Sm
	FF	Foodplain deposits	Fl - Fm - Fo - Po
Río Seco	LS	Laminated sand sheets associated with aeolian deposits	Sl - Fm - Fl - Se
	SB	Sandy bedforms	Gcm - Sm - St - Sp

Table 3. Code of the major architectural elements defined for the Orán Group with their characteristic lithofacies.

http://dx.doi.org/10.5772/intechopen.69985

The deposits of the Anta Formation have been defined as accumulates in a playa lake paleoenvironment, in which the following features have been recognized: sand flats, arid mud-flats, ephemeral saline lakes, and permanent saline lakes [54] (**Figure 11**).

The presence of zeolites (analcime) in the facies of laminated green pelite indicates that it formed in an environment such as alkaline lake in an arid to semiarid condition. In these basins with little or no drainage, evaporation would have increased the alkalinity of the waters and the reaction between the water and volcanic ash falling intermittently into the lake would have caused the zeolitization of the volcanic glass [56, 57].

The provenance of the Metán Subgroup in the area of maximum subsidence (Piedra Blanca, Río Metán, Arroyo González, **Figure 10**) most probably came from the west with sediments of the Salta Group, the Puncoviscana Formation, and granite from the border of Puna as a result of first- or second-order fluvial systems connected to the Calchaquí Basin.

The presence of *Riella* sp. (phylum Bryophyta, class Hepaticae, family Riellaceae) in the Anta Formation suggest that it is the only genus of the class Hepaticae whose present representatives develop in both purely alkaline salt waters and fresh water. Associated with *Pediastrum* sp. and *Phaeceros* sp., they are formed in lacustrine environments and reflect stenohaline conditions that are alkaline and rich in nutrients [54, 58].

4.2. Guanaco Formation (base of Jujuy Subgroup)

The Jujuy Subgroup is widely distributed in the central and southern sector of the Cordillera Oriental and in the Santa Barbara System. It exhibits a general increasing grain-size trend, with cycles of 50–200 m in thickness and lateral extents of tens of kilometers. The cycles represent the progradation of the sediments at times of reduced accommodation space, whereas the cycles of decreasing grain-size represent periods of vertical aggradation associated with greater accommodation space [59] (**Figure 12**).

The basal contact between the Jesús María or older deposits and Guanaco formations is a paraconformity or unconformity and the contact at the top is an unconformity with the Piquete Formation or Quaternary deposits (**Figure 12**).

The Guanaco Formation has garnetiferous glassy tuff layers and is linked to the La Pava-Ramadas Caldera, whose volcanic activity has been dated to 8.73 ± 0.25 Ma (K/Ar; [60]). Records of these tuffs have been found in San Antonio de los Cobres, Lerma Valley, and the valleys of Rio Grande of Jujuy and Juramento-Metán.

The Guanaco Formation was deposited between ~9 and <6.9 Ma [60], near Coronel Moldes, and it has an age of 9.31 ± 0.31 Ma [57]. The thickness of the preserved Guanaco Formation deposits ranges between 0 and 900 m in the Cordillera Oriental and more is more than 2000 m in the Santa Bárbara System (**Figure 12**) [59, 61]. This variation reflects a deformation episode after the formation was deposited and a previous structuring at ~10 Ma [31, 62].

4.2.1. *Facies and depositional architecture of the Guanaco Formation*

The Guanaco Formation is characterized by alluvial fans deposits dominated by: (1) conglomerate and sabulite with channeled bases and an upward fining arrangement, constituting

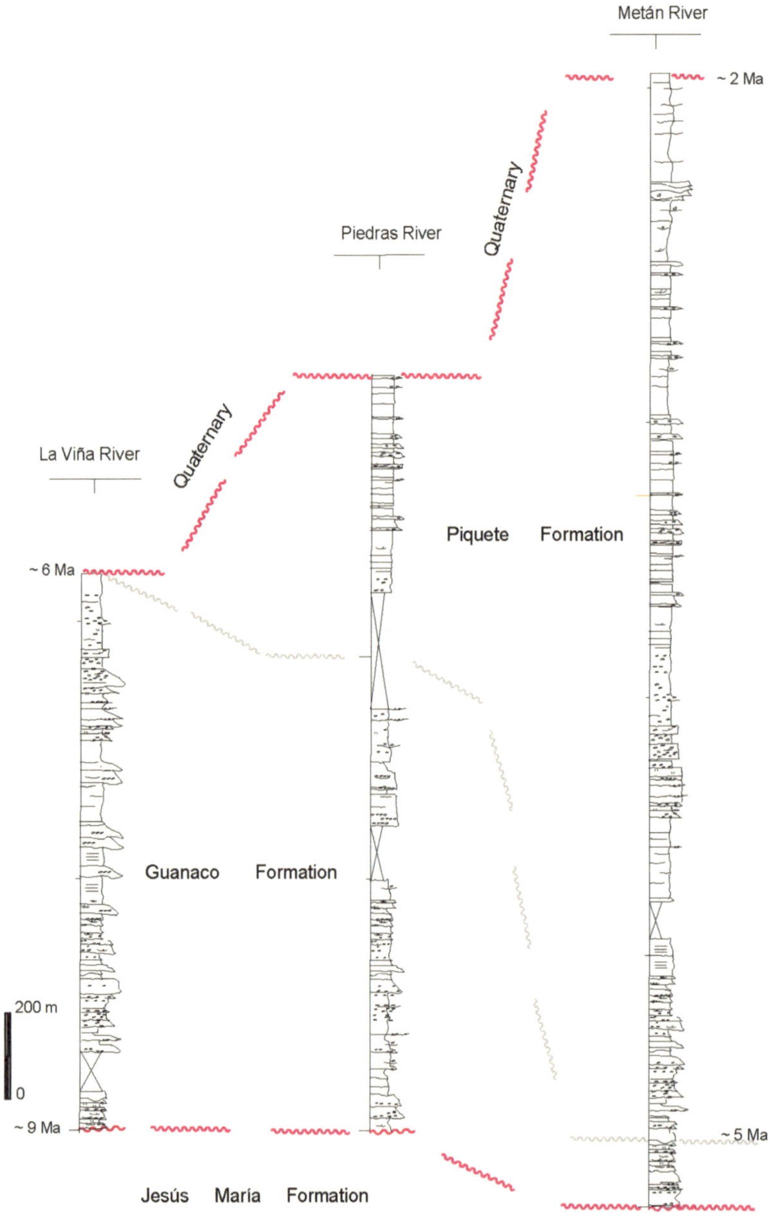

Figure 12. Stratigraphic correlation of the deposits of Jujuy Subgroup (top of Orán Group).

deposits of gravel bars; (2) facies of conglomerates accumulated by hyperconcentrated deposits; and (3) conglomerate sandstone with trough and planar stratification corresponding to lateral bar deposits and dune migration [59] (**Table 3**, **Figure 13a**).

This fluvial system is associated with an alluvial fan paleoenvironment dominated by braided stream. The proximal deposits are located in the western zone (the Lerma Valley) and the

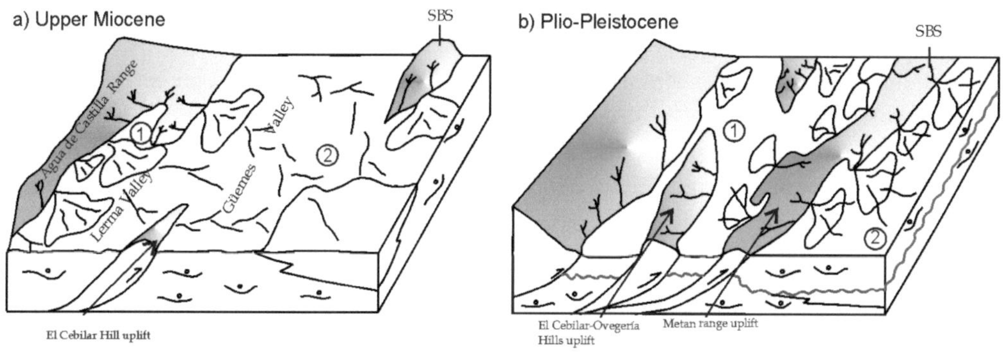

Figure 13. Description of the paleoenvironment with architectural elements illustrated using schematic diagrams of Jujuy Subgroup. a) 1- alluvial fan deposits and, 2- braided fluvial systems deposits of Guanaco Formation, b) 1- and 2- alluvial fan deposits with braided fluvial systems of Piquete Formation.

middle and distal sectors are located in the central and distal zones [59] (**Figures 12** and **13**), which are associated with a large river system.

The conglomerate facies of the Guanaco Formation contain more than 15% high-grade metamorphic clasts (migmatites) and granitoids associated with paleocurrent directions from the west. These characteristics have been interpreted that the provenance sediments is from the eastern edge of the Puna.

The characteristics of the sedimentary paleoenvironment, the provenance data, and the paleocurrent directions from the west suggest that, between ~9 and 6 Ma, the foreland basin evolved independently from the Calchaquí Valley Basin and that the connection of first- and second-order fluvial systems transported material from the eastern edge of the Puna and from the El Toro Lineament.

4.3. Piquete Formation (top of Jujuy Subgroup)

The Piquete Formation is widespread in the Cordillera Oriental, the Santa Bárbara System, and the Sierra de Zapla (**Figures 3** and **12**). The base is characterized by an erosional unconformity or paraconformity on the deposits of the Guanaco Formation in the Santa Bárbara System [63]. In other areas, these units are separated by an angular unconformity, as in the Cordillera Oriental [59, 64] (**Figure 12**).

The measured partial thickness varies from 190 to over 2000 m. This unit features deposits of whitish, fine-grained vitrocrystalline, rhyodacitic to dacitic tuffs with thicknesses of 1.80 to 3 m. The Pliocene deposits of the Piquete Formation accumulated in response to the strong structuring that produced the upheaval of the Subandean ranges, the Santa Bárbara System, and part of the Cordillera Oriental, which also resulted in the formation of intermontane basins, such as the Lerma Valley and the Siancas Valley.

Paleomagnetic studies and ages from fission track dating in apatite from one tuff (Coronel Moldes) in the basal section of this unit yielded an age of 5 Ma [33, 57]. The upper third was dated based on a tuff that yielded an age of 1.3 ± 0.2 Ma [65].

4.3.1. Facies and depositional architecture of the Piquete Formation

The paleoenvironment of the Piquete Formation has been interpreted as relatively small alluvial fans distributed on the flanks of structural depressions and dominated by debris flows (**Table 3**). If these alluvial fans would have been more, had developed in the eastern sector, and away from the thrust fronts, flood plains with small lake systems would have developed [59] (**Figure 13**).

The conglomerates in the Piquete Formation contain slabs of limestone from the Yacoraite Formation, slate from the Precambrian basement of the Puncoviscana Formation, and clasts of reddish sandstone and limestone from the Salta Group. The change in the conglomeratic clast composition from the Guanaco Formation to the Piquete Formation suggests that between ~5 and 2 Ma, thick sediments from the eastern edge of the Puna were trapped in the intermontane Calchaquí Basin [42, 59].

The paleontological content of the Piquete Formation at present is very limited and includes fragmentary remains of vertebrates, notably including abrocomid rodents and plates of Dasypodidae. In the Xibi Xavi river, in the city of Jujuy, complete remains of a glyptodont (*Cranithlastus xibiensis*) and megatherium teeth [66] have been found. Alligatoroid remains from Rosario de la Frontera (south of Salta province), assigned to the species *Caiman latirostris* based on its morphology [67], have also been found.

5. Conclusion

The complexity of tectonic processes controlling the evolution of foreland basins resulted in highly complex basins. The more that is known about these processes and their consequences, the more complex our models become and the more each basin appears to be unique [9].

During the first evolutionary stage of the foreland basin that developed during the middle to upper Eocene in north-western Argentina, the basin had an elongated configuration, was parallel to the Andean uplift, and did not extend to the external sector of the Cordillera Oriental (**Figure 14a**).

The uplift of the margins of the basin and the increase in the relief of the edge of the Puna plateau associated with the Leon Muerto Range are reflected in three depositional sequences that are interpreted to represent three tectonic episodes. Consequently, the main controls over the ephemeral fluvial system were the interactions between tectonics and basin subsidence and the constant arid climatic conditions (**Figure 14a**).

At the beginning of the second evolutionary stage of the foreland basin, an initial elongated depocenter parallel to the orogen developed at ~17 Ma. This depocenter featured the development of playa lake deposits and paleocurrent directions to the north (along the Umbral de Los Gallos). Over time, the depocenter migrated to the eastern edge of the basin by ~15 Ma (**Figure 14b**).

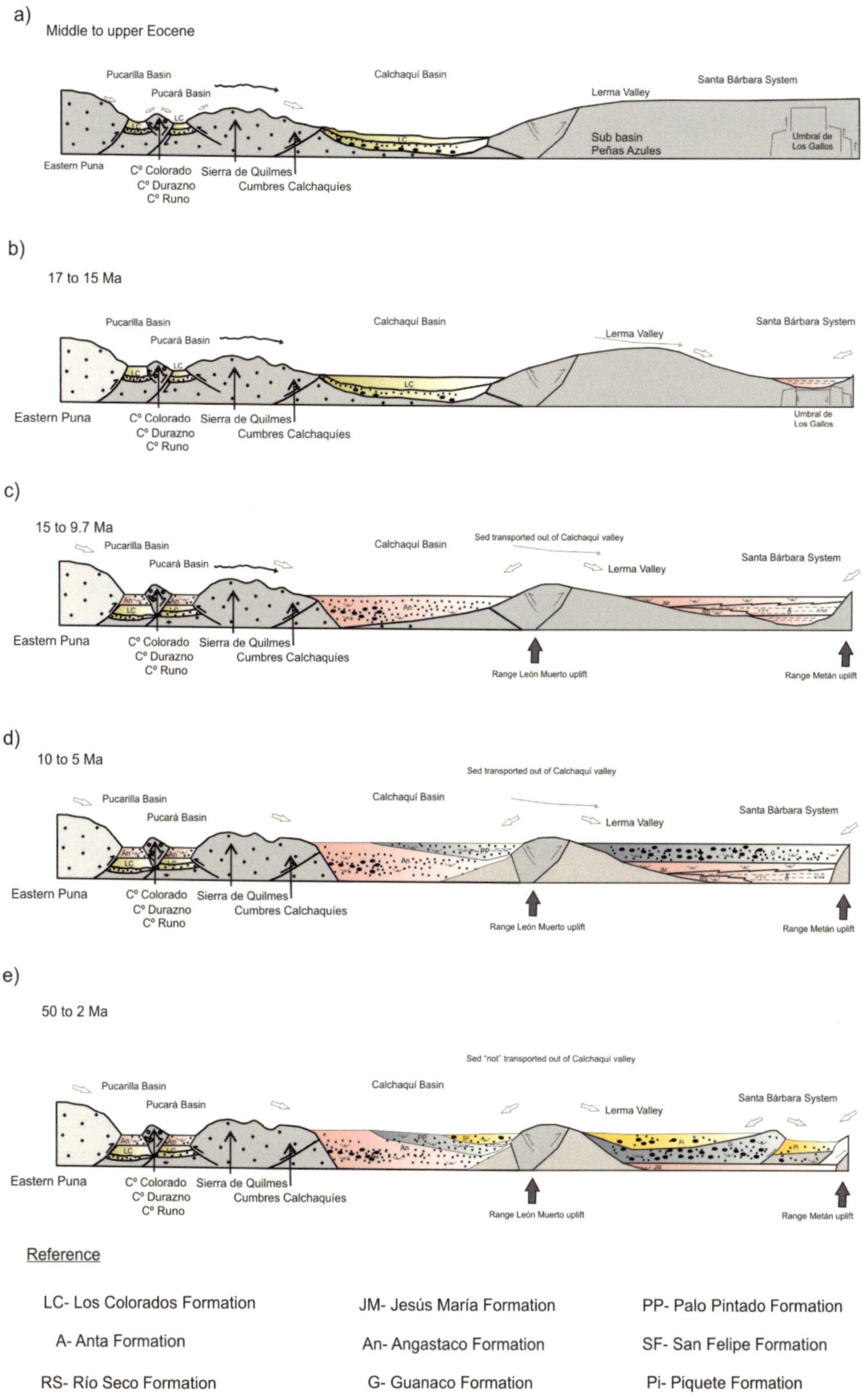

Figure 14. Schematic diagram models for Cenozoic foreland basins showing the evolution of the study area from Eocene (a) to Pliocene (d) time (not to scale).

The second evolutionary stage of the foreland basin (15–10 Ma) featured a new restructuring of the broken foreland basin with the development of thick sedimentary deposits in the Calchaquí basin area that thin toward the eastern basin margin. The same pattern in sedimentation and development is observed in the Metán Subgroup basin to the east (**Figure 14c**).

The contact between the Los Colorados and Angastaco formations is a paraconformity grading into an unconformity. Tectonics and subsidence were the fundamental controls on the evolution of the fluvial style, the deposit thickness, and the paleocurrent variability. The red sandstone clasts (Lumbrera Formation) and gray sandstone clasts (Maíz Gordo Formation) in the Tonco Valley, which are associated with easterly paleocurrents, suggest that the Sierra León Muerto to the east of the Angastaco basin was uplifted (**Figure 14c**).

The Metán Subgroup deposits are interpreted to have accumulated at ~14.9 Ma in an arid paleoenvironment characterized by a sandy ephemeral fluvial system associated with dune fields and playa lake deposits, with sand flats, mud flats, an ephemeral saline lake, and a permanent saline lake and sporadic marine incursions from the south-east (**Figure 14c**). At ~12 Ma, a new basin restructuring event began and the Metán Subgroup deposits began to be eroded.

During the third stage of evolution in the foreland basin (~10–5 Ma), the western part of the basin experienced at least three episodes of tectonic reactivation, which are reflected in variations in the rate of sedimentation in the Palo Pintado Formation. Paleocurrents from the south and south-east indicate tectonic reactivation of the depositional area from the Sierra León Muerto-Sierra Los Colorados (**Figure 14d**).

The Guanaco Formation is characterized by alluvial fans deposits dominated by flowing streams and a braided fluvial system. The sedimentary paleoenvironment, provenance, and paleocurrent data suggest that the foreland basin evolved at a different time and rhythm than the Calchaquí basin, with the connection of first- and second-order river systems transported material from the eastern edge of the Puna and the area of the El Toro Lineament (**Figure 14d**).

The San Felipe Formation is characterized by braided fluvial fan and a shallow gravel-braided fluvial system. The provenance and abundant clasts in different levels of the Salta Group and the association with paleocurrents from the north-east, east, and south-east suggest a reactivation of the Sierra León Muerto and the Sierra Los Colorados in the depositional area (**Figure 14e**).

The Piquete Formation lies in marked unconformity over the deposits of the Guanaco Formation or older deposits. They accumulated as a series of alluvial fans of limited dimensions and are distributed on the flanks of structural depressions and dominated by debris flows. The composition of the clasts of conglomerates from the Piquete Formation suggests that between ~5 and 2 Ma, the basin was isolated from the basin of the San Felipe Formation (**Figure 14d**).

Acknowledgements

This research was funded by the projects: AGENCIA (PICT-2012-1984, PICT 2014-3654), UNJu (SECTER 08/E036 and 08/E037), UNSa (CI-UNSa 2287), and SuRfAce processes, TEctonics and Georesources: The Andean foreland basin of Argentina (StRaTEGy).

Author details

Claudia Inés Galli[1,3]*, Ricardo Narciso Alonso[2] and Lidia Beatriz Coira[3]

*Address all correspondence to: claudiagalli@fibertel.com.ar

1 INECOA-Facultad de Ingeniería, Universidad Nacional de Jujuy, Facultad de Ciencias Naturales, Universidad Nacional de Salta, Argentina

2 CEGA-CONICET, Universidad Nacional de Salta, Salta, Argentina

3 INECOA-CONICET, Instituto de Geología y Minería, S.S. de Jujuy, Argentina

References

[1] Jordan TE. Thrust loads and foreland basin evolution, Cretaceus, western United States. American Association of Petroleum Geologists Bulletin. 1981;**65**:2506-2520

[2] Beaumont C. Foreland basins. Geophysical Journal of the Riyal Astronomical Society. 1981;**65**:291-329. DOI: 10.1111/j.1365-246x.1981.tb02715.x

[3] Jordan TE. Retroarc foreland and related basins. In: Busby CJ, Ingersoll RV, editors. Tectonics of Sedimentary Basins. 1995th ed. Blackwell Science Inc. Cambridge, MA; 1995. pp. 331-362

[4] DeCelles PG, Giles KA. Foreland basin systems. Basin Research. 1996;**8**:105-123. DOI: 1365-2117.1996.01491.x

[5] Catuneanu O. Retroach foreland systems: Evolution through time. Journal of African Earth Science. 2004;**38**:225-242. DOI: 10.1016/j.jafrearsci.2004.01.004

[6] DeCelles PG, Carrapa B, Horton B, Gehrels GE. Cenozoic foreland basin system in the central Andes north-western Argentina: Implications for Andean geodynamics and modes of deformation. Tectonics. 2011;**30**:TC6013. DOI: 10.1029/2011TC002948

[7] Carrapa B, Reyes-Bywater S, Safipour R, Sobel ER, Schoenbohm LM, DeCelles PG, Reiners PW, Stockli D. The effect of inhereted paleotopography on exhumation of the Central Andes of NW Argentina. Geological Society of America Bulletin. 2014;**30844**:1-12. DOI: 10.1130/B30844.1

[8] Price RA. Large-scale gravitational flow of supracrustal rocks, Southern Cabadian Rockies. In: DeJong KA, Scholten R, editors. Gravity and Tectonics. 1973rd ed. New York: John Wiley and Sons; 1973. pp. 491-502

[9] Ingersoll RV. Tectonics of sedimentary basins, with revised nomenclature. In: Busby C, Azor Pérez A, editors. Tectonics of Sedimentary Basins Recent Advances. 2012th ed. Blackwell Publishing Ltd. Oxford, UK; 2012. pp. 3-46. DOI: 10.1002/9781444347166.ch1

[10] Strecker MR, Hilley GE, Bookhagen B, Sobel ER. Structural, geomorphic and depositional characteristics of contiguous and broken foreland basins: Examples from the eastern flanks of the central Andes Bolivia and NW Argentina. In: Busby CA, Azor Perez A, editors. Tectonics of Sedimentary Basins Recent Advances. 2012th ed. USA: Blackwell Publishing Ltd; 2012. pp. 508-522. DOI: 10.1002/9781444347166.ch25

[11] Jordan TE, Alonso RN. Cenozoic stratigraphy and basin tectonics of the Andes Mountains, 20°-28° South latitude. American Association of Petroleum Geologist Bulletin. 1987; **71**:49-64

[12] Ramos VA. Evolución tectónica de la Argentina. In: Caminos R, editors. Geología Argentina. 1999th ed. Buenos Aires: Instituto de Geología y Recursos Minerales - SEGEMAR; 1999. pp. 715-784

[13] Hongn F, del Papa C, Powell J, Petrinovic I, Mon R, Deraco V. Middle Eocene deformation and sedimentation in the Puna-Eastern Cordillera transition (23°-26° S): control by preexisting heterogeneities on the pattern of initial Andean shortening. Geology. 2007;**35**:271-274. DOI: 10.1130/G23189A.1

[14] Hongn F, Mon R, Petrinivic I, del Papa C, Powell J. Inversión y ractivación tectónica cretácica en el noroeste argentino: Influencia de las heterogeneidades del basamento Neoproterozoico-Paleozoico inferior. Revista de la Asociación Geológica Argentina. 2010;**66**:38-53

[15] Toselli AJ. Metamorfismo del Ciclo Pampeano. In: Aceñolaza FG, Miller H, Toselli AJ, editors. El Ciclo Pampeano en el Noroeste Argentino. 1990th ed. Tucuman, Argentina: INSUGEO; 1990. pp. 181-197

[16] Turner JC, Mon R. Cordillera oriental. In: II Simposio Geología Regional. 1979th ed. Córdoba, Argentina: Academia Nacional De Ciencias de Córdoba; 1979. pp. 163-169

[17] Aceñolaza FG, Toselli A. Consideraciones estratigráficas y tectónicas sobre el Paleozoico inferior del Noroeste Argentino. In: 2° Congreso Latinoamericano de Geología. Caracas, Venezuela: 1976. pp. 755-763

[18] Toselli AJ, López JP, Sardi FG. El basamento metamórfico en Cumbres Calchaquíes Noroccdentales, Aconquija, Ambato y Ancasti: Sierras Pampeana. In: Gonzáles Bonorino G, Omarini R, Viramonte J, editors. Geología del Noroeste Argentino, Relatorio. 9° Congreso Geológico Argentino. Salta: 1999. pp. 73-77.

[19] Turner JC. Estratigrafía de la Sierra de Santa Victoria y adyacencias. Academia Nacional de Ciencias de Córdoba Boletín. 1960;**41**:163-196

[20] Starck D. Silurian-Jurassic stratigraphy and evolution of north-western Argentina. In: Tankard AJ, Suarez Soruco R, Welsink HJ, editors. Petroleum of South America. 1st ed. American Association of Petroleum Geologists Memoir. USA; 1995. pp. 137-184

[21] Brackebusch L. Estudio sobre la Formación Petrolífera de Jujuy. Academia Nacional de Ciencias de Córdoba Boletín. Argentina : 1883;**2**:137-184

[22] Reyes FC, Salfity JA. Consideraciones sobre la estratigrafía del Cretácico (Subgrupo Pirgua) del noroeste argentino. In: Actas 5° Congreso Geológico Argentino. Carlos Paz, Córdoba: 1973. pp. 355-385

[23] Moreno JA. Estratigrafía y paleogeografía del Cretácico superior en la cuenca del noroeste Argentino, con especial mención de los Subgrupo Balbuena y Santa Bárbara. Revista de la Asociación Geológica Argentina. 1970;**25**:9-44

[24] Cazau L, Cellini N, Oliver G, J. El Subgrupo Santa Bárbara (Grupo Salta) en la porción oriental de las provincias de Salta y Jujuy. In: 6° Congreso Geológico Argentino. Bahía Blanca: 1976. pp. 341-355

[25] Pascual R, Vucetich MG, Fernández J. Los primeros mamíferos (Notoungulata, Henricorbornidae) de la Formación Mealla, su importancia filogenética, taxonómica y cronológica. Ameghiniana. 1978;**15**:366-390

[26] Bond M, López G. El primer Notohippidae (Mammalia, Notoungulata) de la Formación Lumbrera (Grupo Salta) del noroeste argentino. Consideraciones sobre la familia Notohippidae. Ameghiniana. 1993;**30**:59-68

[27] Babot MJ, Powell JE, De Muizon C. *Callistoe vincei*, a new Proborrhyaenidae (Borhyaenoidea, Metatheria, Mammalia) from the Early Eocene of Argentina. Geobios. 2002;**35**: 615-629

[28] Pascual R. Nuevos y singulares tipos ecológicos de marsupiales extinguidos de América del Sur (Paleoceno tardío - Eoceno Temprano) del noroeste argentino. 2° Congreso Argentino de Paleontología y Bioestratigrafía, 1° Congreso Latinoamericano de Paleontología. 1980: 151-173

[29] Starck D, Vergani G. Desarrollo tecto-sedimentario del Cenozoico en el sur de la provincia de Salta, Argentina. 13° Congreso Geológico Argentino. 1996:433-452

[30] del Papa C. Estratigrafía y paleoambientes de la Formación Lumbrera, Grupo salta, Noroeste Argentino. Revista de la Asociación geológica Argentina. 2006;**61**:313-327

[31] del Papa C, Kirschbaum A, Powell J, Brod A, Hongn F, Pimentel M. Sedimentological, geochemical and paleontological insights applied to continental omission surfaces: A new approach for reconstructing an Eocene foreland basin in NW Argentina. Journal of South America Earth Science. 2010;**29**:327-345. DOI: 10.1016/j.jsames.2009.06.004

[32] Russo A. La estratigrafía terciaria en el noroeste argentino. 5° Congreso Geológico Argentino, Actas. 1972;**1**:1-14

[33] del Papa C, Hongn F, Petrinovich I, Domingues R. Evidencias de deformación pre-miocena asociada al antepaís andino en la Cordillera Oriental (24° 35′ S- 66° 12′ O). Nota Breve. Revista de la Asociación Geológica Argentina. 2004;**59**:506-509

[34] Payrola Bosio P, del Papa C, Hongn F, Powell J. Estratigrafía del Valle de Luracatao (Valle Calchaquí, Noroeste Argentino): Nueva propuesta. Revista de la Asociación Geológica Argentina. 2010;**67**:309-318

[35] Vail P, Audemard F, Bowman S, Eisner P, Pérez Cruz C. The stratigraphic signatures of tectonics, eustasy and sedimentology - An overview. In: Einsele G, editor. Cycles and Events in Stratigraphy. 1992nd ed. Berlin: Springer; 1992. pp. 617-659

[36] Vail PR, Mitchum EM, Thompson S. Seismic stratigraphy and global changes of sea level, part 4. In: Payton CA, editors. Seismic Stratigraphy - Aplication to Hidrocarbon Exploration. 1977th ed. Memoir American Association Petroleum Geologist. USA; 1977;**26**:83-97

[37] Carrapa B, Trimble J, Stockli D. Patterns and timing of exhumation and deformation in the Eastern Cordillera of NW Argentina revealed by (U-Th)/He thermochronology. Tectonics. 2011;**30**:TC3003. DOI: 10.1029/2010TC002707

[38] Galli CI, Reynolds J. Evolución paleoambiental del Grupo Payogastilla (Eoceno-Plioceno) en el Valle Calchaquí - Tonco, provincia de Salta, Argentina. In: Marquillas RA, Sánchez MC, Salfity JA, editors. Aportes sedimentológicos a la geología del noroeste argentino. 1st ed. Salta, Argentina: SCS Publisher; 2012. pp. 67-80. ISBN 978-987-26890-1-8

[39] del Papa C, Hongn F, Powell J, Payrola Bosio P, Do Campo M, Strecker MR, Petrinovic I, Schmitt AK, Pereyra R. Moddle Eocene-Oligocene broken-foreland evolution in the Andean Calchaquí Valley, NW Argentina: Insights from stratigraphic, structural and provenance studies. Basin Research. 2013;**25**:1-20. DOI: 10.1111/bre.12018

[40] Catuneanu O, Abreu V, Bhattacharya JP, Blum MD, Dalrymple RW, Eriksson PG, et al. Towards the standardization of Sequence Stratigraphy. Earth and Atmospheric Sciences. 2009. Paper 238 http://digitalcommons.unl.edu/geosciencefacpub/238

[41] Iaffa DN, Sabat F, Bello D, Ferrer O, Mon R, Gutierrez A. Tectonic inversion in a segmented foreland basin from extensional to piggy back settings: The Tucumán basin in NW Argentina. Journal of American Earth Sciences. 2011;**31**:457-474. DOI: 10.1016/j.jsames.2011.02.009

[42] Hain M, Strecker M, Bookhagen B, Alonso RN, Pingel H, Schmitt A. Neogene to Quaternary broken foreland formation and sedimentation dynamics in the Andes of north-west Argentina (25°S). Tectonics. 2011;**30**:TC2006. DOI: 10.1029/2010TC002703

[43] Galli CI, Coira LB, Alonso RN, Reynolds J, Matteini M, Hauser N. Tectonic controls on the evolution of the Andean Cenozoic foreland basin: Evidence from fluvial system variations in the Payogastilla Group, in the Calchaquí, Tonco and Amblayo valleys, NW Argentina. Journal of South America Earth Sciences. 2014a;**52**:234-259. DOI: 10.1016/j.jsames.2014.03.003

[44] Vezzoli L, Acocella V, Omarini R, Mazzuoli R. Miocene sedimentation, volcanic and deformation in the Eastern Cordillera (24°30′ S, NW Argentina): Tracking the evolution of the foreland basin of the Central Andes. Basin Research. 2012;**24**:1-27. DOI: 10.1111/j.1365-2117.2012.00547.x

[45] Galli CI, Ramirez A, Barrientos C, Reynolds J, Viramonte JG, Idleman B. Estudio de proveniencia de los depósitos del Grupo Payogastilla (Mioceno Medio-Superior) aflorantes en el río Calchaquí, provincia de Salta, Argentina. 17° Congreso Geologico Argentino, Jujuy. 2008;**1**:353-354

[46] Coutand I, Carrapa B, Deeken A, Schmitt AK, Sobel E, Strecker M. Orogenic plateau formation and lateral growth of compressional basins and ranges: Insights from sandstone petrography and detrital apatite fission-track thermochronology in the Angastaco Basin, NW Argentina. Basin Research. 2006;**18**:1-26. DOI: 10.1111/j.1365-2117.2006.00286.x

[47] Bywater-Reyes S, Carrapa B, Clementz M, Clementz M, Schoenbohm L. Effect of late Cenozoic aridification on sedimentation in the Eastern Cordillera of north-west Argentina (Angastaco basin). Geology. 2010;**38**:235-238. DOI: 10.1130/G30532.1

[48] Galli CI, Anzotegui LM, Horn MY, Morton LS. Paleoambiente y paleocomunidades de la Formación Palo Pintado (Mioceno-Plioceno, Provincia de Salta, Argentina. Revista Mexicana de Ciencias Geológicas. 2011b;**28**:161-174. ISSN 2007-2902

[49] Starck D, Anzotegui L. The late Miocene climatic change persistence of a climatic signal through the orogenic stratigraphic record in north-western of Argentina. Journal of South America Earth Sciences. 2001;**14**:763-774. DOI: 10.1016/S0895-9811(01)00066-9

[50] Galli CI, Vides ME, Flores P. Sedimentological and climatic control in the clay minerals distribution in the fluvial deposits of the Palo Pintado Formation (upper Miocene), Salta Province, Argentina. 18° International Sedimentological Congress "Sedimentology at the Food of the Andes". 2010:376

[51] Bona P, Starck D, Galli CI, Gasparini Z, Reguero M. Caiman cf. Latirostris (Alligatoridae, caimaninae) in the late Miocene Palo Pintado Formation, Salta province, Argentina: ¨Paleogeographic and Paleoenvironmental consideration. Ameghiniana. 2014;**51**:26-36

[52] Galli CI, Coira LB, Alonso RN, Matteini M, Hauser N. El Grupo Payogastilla (Cenozoico) en los valles Calchaquí, Tonco y Amblayo, provincia de Salta, Argentina. Acta Geológica Lilloana, Tucumán, Fundación Miguel Lillo. 2015;**26**(1852-6217):30-52

[53] Salfity JA, Monaldi CR, Marquillas RA, González RE. La inversión tectónica del Umbral de Los Gallos en la cuenca del Grupo Salta durante la Fase Incaica. 12° Congreso Geológico Argentino y 2° Congreso de Exploración de Hidrocarburos. 1993;**3**:200-210

[54] Galli CI. Estratigrafía y sedimentología del Subgrupo Metán (Grupo Orán, Terciario), provincia de Salta, Argentina [thesis]. 1995. p. 109

[55] Reynolds J, Idleman B, Hernández R, Naeser CW. Preliminary chronostratigraphic constraints on Neogene tectonic activity in the Eastern Cordillera and Santa Bárbara System, Salta province, NW Argentina. In: Geological Society of America, Abstracts with Programs; Seattle, WA. 1994:503

[56] Galli CI, Hernández R, Reynolds J. Análisis estratigráfico del Subgrupo Metán (Grupo Orán), en el río Piedras, departamento Metán, Salta, Argentina. Boletín de Informaciones Petroleras. 1996;**12**:99-107

[57] Reynolds J, Galli CI, Hernández R, Idleman B, Kotila J, Hilliard R, Naeser CW. Neogene foreland uplift history in the flat subduction region, transition zone and normal subduction region, NW Argentina. In: Geological Society of America. Abstract with Programs; Reno, NV. 2000:504-505

[58] Quattrocchio M, Durango de Cabrera J, Galli CI. Formación Anta (MIOceno temprano-medio), Subgrupo Metán (Grupo Orán), en el río Piedras, provincia de Salta. Revista de la Asociación Geológica Argentina. 2003;**58**:117-127

[59] González Villa RE. El Subgrupo Jujuy (Neógeno) entre los 24°-26° LS y 64°-66° LO, tramo centro austral de la cadena subandina Argentina, provincias de Salta y Jujuy [thesis]. 2002.p. 405

[60] Viramonte JG, Reynolds J, del Papa C, Disalvo A. The Corte Blanco garnetiferous tuff: A distinctive late Miocene marker bed in northwestern Argentina applied to magnetic polarity stratigraphy in the Río Yacones, Salta Province. Earth Planetary Science Letter. 1994;**121**:519-531. DOI: 10.1016/0012-821x(94)90088-4

[61] Gebhard J, Giudici AR, Oliver Gascón J. Geología de la comarca entre el río Juramento y arroyo Las Tortugas, provincia de Salta y Jujuy, República Argentina. Revista de la Asociación Geológica Argentina. 1974;**29**:359-375

[62] Strecker M, Alonso RN, Bookhagen B, Carrapa B, Coutand I, Hain M, Hilley, et al. Does the topographic distribution of the central Andean Puna Plateau result from climatic or geodynamic processes? Geology. 2009;**37**:643-646. DOI: 10.1130/G25545A.1

[63] Cristallini E, Cominguez AH, Ramos VA. Deep structure of the Metán-Guachipas region: Tectonic inversion in North-western Argentina. Journal of South American Earth Sciences. 1997;**10**:403-421

[64] Carrera N, Muñoz J. Thrusting evolution in the southern Cordillera Oriental (northern Argentina Andes): Constraints from growth strata. Tectonophysics. 2008;**459**:107-122. DOI: 10.1016/j.tecto.2007.11.068

[65] Malamud BD, Jordan T, Alonso RN, Gallardo E, González R. Four new Quaternary ash and tuff ages, Lerma valley, NW Argentina. American Geophysical Union, Fall Meeting. 1995;**V22b**:12

[66] Arias J, Alonso RN, Malanca S. Un gliptodontoideo de la Formación Piquete (Grupo Orán), Provincia de Jujuy, República Argentina. Revista del Instituto de Ciencias Geológicas. Jujuy, Argentina. 1978;**3**:175-188.

[67] Barrios F. Presencia de *Caiman latirostris* (Daudin, 1802) (Crocodylia Alligatoridae) en la Formación Piquete (Plioceno-Pleistoceno Temprano) de la Provincia de Salta, Argentina: Implicancias Paleoambientales y Sistemáticas. Ameghiniana. 2013;**50**:522-534

[68] Salfity JA, Monaldi CR. Hoja Geológica 2566-IV, Metán. Programa Nacional de Cartas Geológicas de la Repúblical Argentina, SEGEMAR. 2006;**319**:74

[69] Hongn F, Seggiaro R. Hoja Geológica 2566-III, Cachi. Programa Nacional de Cartas Geológicas de la República Argentina. 2001;**Boletín**:74